29.—

FOURIER TRANSFORM INFRARED SPECTROSCOPY

TECHNIQUES USING FOURIER TRANSFORM INTERFEROMETRY

Volume 3

CONTRIBUTORS

C. L. ANGELL

JOHN R. FERRARO

R. H. HALL

GARY HORLICK

K. KRISHNAN

LAURENCE A. NAFIE

D. WARREN VIDRINE

W. K. YUEN

FOURIER TRANSFORM INFRARED SPECTROSCOPY

TECHNIQUES USING FOURIER TRANSFORM INTERFEROMETRY

Edited by

JOHN R. FERRARO
Department of Chemistry
Loyola University
Chicago, Illinois

LOUIS J. BASILE
Chemistry Division
Argonne National
 Laboratory
Argonne, Illinois

VOLUME 3

 1982

ACADEMIC PRESS
A Subsidiary of Harcourt Brace Jovanovich, Publishers
New York   London
Paris   San Diego   San Francisco   São Paulo   Sydney   Tokyo   Toronto

ACADEMIC PRESS, INC.
111 Fifth Avenue, New York, New York 10003

United Kingdom Edition published by
ACADEMIC PRESS, INC. (LONDON) LTD.
24/28 Oval Road, London NW1 7DX

Library of Congress Cataloging in Publication Data
Main entry under title:

Fourier transform infrared spectroscopy.

   Includes bibliographies and indexes.
   1. Infra-red spectrometry. 2. Fourier transform
spectroscopy. I. Ferraro, John R., Date.
II. Basile, Louis J.
QD96.I5F68   535.8'42   77-75571
ISBN 0-12-254103-0 (v. 3)   AACR1

PRINTED IN THE UNITED STATES OF AMERICA

82 83 84 85   9 8 7 6 5 4 3 2 1

CONTENTS

1 FOURIER TRANSFORM INFRARED SPECTROSCOPY IN THE STUDY OF CATALYSTS
C. L. ANGELL

2 ATOMIC EMISSION SPECTROCHEMICAL MEASUREMENTS WITH A FOURIER TRANSFORM SPECTROMETER
GARY HORLICK, R. H. HALL, AND W. K. YUEN

3 DOUBLE MODULATION FOURIER TRANSFORM SPECTROSCOPY
LAURENCE A. NAFIE AND D. WARREN VIDRINE

4 PHOTOACOUSTIC FOURIER TRANSFORM INFRARED SPECTROSCOPY OF SOLIDS AND LIQUIDS
D. WARREN VIDRINE

5 TECHNIQUES USED IN FOURIER TRANSFORM INFRARED SPECTROSCOPY
K. KRISHNAN AND JOHN R. FERRARO

LIST OF CONTRIBUTORS

Numbers in parentheses indicate the pages on which the authors' contributions begin.

C. L. ANGELL (1), Union Carbide Corporation, Tarrytown, New York 10591

JOHN R. FERRARO (149), Department of Chemistry, Loyola University, Chicago, Illinois 60626

R. H. HALL (37), Analytical Research, Syncrude Canada Ltd., Edmonton, Alberta T6G 2G2, Canada

GARY HORLICK (37), Department of Chemistry, University of Alberta, Edmonton, Alberta T6G 2G2, Canada

K. KRISHNAN (149), Digilab Division, Bio Rad Laboratories, Cambridge, Massachusetts 02139

LAURENCE A. NAFIE (83), Department of Chemistry, Syracuse University, Syracuse, New York 13210

D. WARREN VIDRINE (83, 125), Nicolet Instrument Corporation, Madison, Wisconsin 53711

W. K. YUEN (37), Saskatchewan Research Council, Saskatoon, Saskatchewan S7N 0X1, Canada

PREFACE

The continued interest in Fourier transform interferometry (FT-IR) has motivated the editors and publishers to publish a third volume in this treatise on the subject. Applications have surfaced using a Fourier transform interferometer for purposes other than infrared studies. In addition, several new linkups between other techniques and FT-IR have occurred. As a consequence of these developments we have prepared Volume 3.

We stress the fact that it has not been our intention in any of the three volumes to indicate that only FT-IR can solve the various problems discussed. Certainly, computer-dispersive infrared spectroscopy is alive and well. In effect, the chemist is blessed in that he has two tools to choose from.

Volume 3 consists of five chapters prepared by people who are well informed in their respective areas. Surface studies of catalytic surfaces are summarized in Chapter 1. Spectrochemical analysis utilizing an FT interferometer is discussed in Chapter 2. New extensions of FT-IR linking it with vibrational circular dichroism and photoacoustic spectroscopy are presented in Chapters 3 and 4. Chapter 5 presents an update, review, evaluation, and summary of many of the techniques that have been used in FT-IR spectroscopy.

It is our hope that the volumes have served, and will continue to serve, as both a stimulus to the scientific community and an example of the viability of the FT-IR technique for the solution of analytical as well as fundamental chemical problems.

1

FOURIER TRANSFORM INFRARED SPECTROSCOPY IN THE STUDY OF CATALYSTS

C. L. Angell

Union Carbide Corporation
Tarrytown, New York

I. INTRODUCTION

Infrared spectroscopy is a well-accepted technique for the study of catalysts and surface reactions. In the mid-1950s the first articles on this subject appeared, and the success of the method can probably be estimated by the approximately one thousand articles that have subsequently appeared. Two books (Little, 1966; Hair, 1967) and several major review articles (Little, 1971; Pritchard and Catterick, 1976; Delgass *et al.*, 1979; Bell, 1980) survey the field, although an up-to-date review is much needed.

The success of infrared spectroscopy can be ascribed to the fact that it is one of the few methods that allow direct examination of adsorbed molecules on solid surfaces. The systems studied include amorphous solids like silica and silica–alumina and crystalline solids like metal oxides, molecular sieves, and metal dispersions on most of the above as supports. The surface areas of these solids vary from a few to ~ 1000 m^2 g^{-1}.

These studies fall into several categories: (a) the nature of the catalysts themselves, e.g., types, quantities, and acidities of hydroxyl groups pres-

FOURIER TRANSFORM
INFRARED SPECTROSCOPY, VOL. 3
Copyright © 1982 by Academic Press, Inc.
All rights of reproduction in any form reserved.
ISBN 0-12-254103-0

ent on the catalyst surface; (b) the nature of the adsorbed species, usually deduced from the shift in frequency of a band on adsorption, e.g., carbon monoxide or carbon dioxide adsorbed on cations in molecular sieves; (c) the strength of adsorption, studied by the removal of characteristic bands of some adsorbed species on pumping at various higher temperatures; (d) the quantity of adsorbed species, usually from band intensities after extinction coefficients of the adsorbed species are established by comparison with quantitative gravimetric measurements; (e) the formation of some new species by surface reaction, e.g., isomerization of olefins on acidic catalysts, formation of carbonate-type species by the reaction of carbon dioxide with oxide catalysts, or the appearance of a saturated species on the interaction of olefins with metal surfaces; and (f) sometimes, the kinetics of these surface reactions. These studies, in many cases, were followed by force constant calculations and interpretation of the mechanism of reactions at the surface.

The purpose of this chapter is to evaluate the effect of the advent of commercially available Fourier transform infrared spectrometers on the well-established method of surface and catalyst studies by infrared spectroscopy. It would seem superfluous to refer in this third volume of a treatise dealing with Fourier transform infrared spectroscopy to the two advantages of interferometers: the Fellgett advantage and the Jacquinot advantage. Some portion of these advantages is very often lost in actual experimental situations. However, some very useful capabilities of FT-IR systems will be discussed: very high signal-to-noise ratio leading to high sensitivity, rapid data acquisition, reduced heating effect of the beam, and the data-handling capabilities provided by a dedicated computer. The latter is not unique to interferometers but would apply to dispersive instruments equipped with a computer. It should be noted that although a large number of papers dealing with infrared studies of catalytic systems have appeared in the past few years, this chapter will discuss only those articles that used FT-IR spectroscopy.

II. EXPERIMENTAL CELLS

A great number and variety of cells have been described in the literature [for some, see Little (1966), p. 31]. All have a common purpose; i.e., to place a sample of a solid material to be studied into a closed system where it can be heat-treated either in situ or in a special furnace section, where it can be exposed to a known pressure of some gaseous material and where its infrared spectrum can be obtained. In 1965 we described a simple cell (Angell and Schaffer, 1965) with the following advantages (Fig. 1): the sample can be moved, with the help of an external magnet, from a

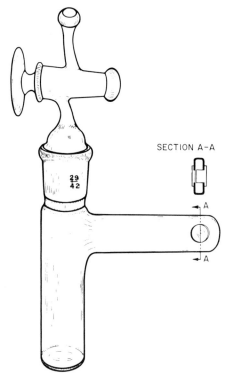

SECTION A-A

Fig. 1.   Short-pathlength infrared cell. Reprinted with permission from Angell and Schaffer (1965), *J. Phys. Chem.* **69,** 3463. Copyright 1965 American Chemical Society.

heating section to an examination section between infrared transmitting windows (a common property of many of the other cells described), the pathlength between the windows in the infrared beam is quite small to minimize the unwanted contribution from the gas-phase spectrum, and the whole cell can be constructed easily and at a low price, making it possible to have and operate a number of cells. In the time since the publication of the description of the cell, this last advantage has proved of great help. Many times when a series of samples had to be studied, there were several cells on the vacuum system with samples being activated, one cell in the spectrometer being scanned, and several others standing, sometimes for days, with the samples exposed to some gas as part of prolonged time studies. Several modifications have been made on the original design: (a) a few cells have been constructed from quartz to allow treatment at higher temperatures; (b) since, in most cases, the materials (silica, zeolites, etc.) studied do not transmit in the region below 1200 cm^{-1}, the original NaCl windows were replaced by CaF$_2$ windows for routine work, although in some cases CsBr windows were used; (c) in some studies, it was necessary to add known amounts of water to the cell already containing several hundred torrs of some gas. For this purpose, a very small side arm with a rubber septum was added to some of the cells and the water added in the liquid form with a syringe.

The only disadvantage of this cell is that it is not possible to take spectra with the sample at higher than room temperatures. There are some cells described in the literature where this is possible, but all of these are quite elaborate and cumbersome, and it is hard to imagine that any laboratory would build more than one of them.

Another type of cell used in FT-IR work has been described by Ceckiewicz and Galuszka (1976). It incorporates the feature that the sample can be moved from a heating region to a spectral examination region, but it does not have the feature of a short pathlength in the beam. This does not matter in the experiments described, since they are dealing with strong chemisorption where all the gas phase is removed by adsorption on the solid. The most important feature of the cell is that it is completely free from any metallic parts; even the sample holder is made of glass. This could be of major importance when there is a possibility that some catalytic reaction could occur on metal surfaces.

Young and Sheppard (1967) described a double-beam infrared cell, which was originally designed to be used in a Grubb–Parsons double-

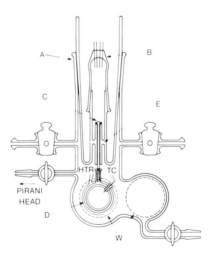

Fig. 2. Double-beam infrared cell (Young and Sheppard, 1967): A, Ground glass joint; B, tungsten–glass seals for leads; C, stainless steel rod to support disk holder; D, stainless steel disk holder; W, calcium fluoride windows; TC, thermocouple; and HTR, Nichrome heating wire.

beam grating instrument but was later also used in their FT-IR work. The cell body is made of Pyrex with calcium fluoride windows and an optical pathlength of 10 cm. The sample can be heated in situ to 350°C (Fig. 2). The cell extends into both beams of the spectrometer; since the gas pressure is uniform throughout the cell, at room temperature the vapor-phase spectrum can be accurately balanced in the two beams of the spectrometer, thus permitting direct recording of the spectrum of the adsorbed phase. The authors reported, however, that since only the sample side of the cell could be heated, temperature differentials were set up at high sample temperature with the result that double-beam cancellation of the vapor-phase spectra became imperfect at pressures over 10 Torr at temperatures above ~ 50°C.

III. RESULTS

A. Hydroxyl Stretching Region

A large amount of effort by many researchers has been expended on the study of the hydroxyl stretching bands of a variety of inorganic support materials. A considerable amount of information, qualitative as well as quantitative, has been accumulated about the nature of surface hydroxyl groups and their behavior in interaction with adsorbed molecules. In the vast majority of the systems studied (silicas, silica-supported systems, zeolites), there is very little transmission in the hydroxyl stretching region because of light scattering due to the particles of the material being approximately the same size as the wavelength of the light in this region (2–5 μm). Although in a few cases it was possible to separate small particles from the material (Angell and Schaffer, 1965), in most cases it was necessary to work with samples as they were. While there is little transmission in the hydroxyl stretching region, in most cases, other parts of the spectrum (between 2000 and 1200 cm^{-1}) show high transmission (as much as 60–70%). The problem, therefore, can be stated as follows: How can spectra of samples be obtained that have very low transmittance only in regions of interest and high transmittance in other regions? In the case of dispersive instruments, the solution is quite simple. Since this type of instrument only looks at a small part of the spectrum at any given time, one can increase the gain, open the slits, and decrease the scanning speed. In this way, one can obtain good-quality spectra at the expense of much longer scanning times.

However, in the case of the interferometer, the situation is very different. Since the interferometer looks at the total spectrum at all times, there is always plenty of energy going through the sample. Therefore, one cannot increase the gain without overloading the detector or overflowing the analog-to-digital converter (ADC). In our laboratory, we have developed several methods to deal with the situation, some of which deal with the general improvement of the quality of the spectra. These are (a) stored phase, (b) changing mirror velocity, (c) cooled detectors, (d) electronic filtering, (e) optical filtering, and (f) smoothing.

a. Stored Phase. It was found that when the sample and background interferograms were treated individually in the normal manner (apodize, phase calculate, transform, and phase correct, before ratioing), the resulting spectrum usually showed a large, broad band around 3500–3200 cm^{-1} (Fig. 3), which clearly is an artifact since the spectrum of the same sample

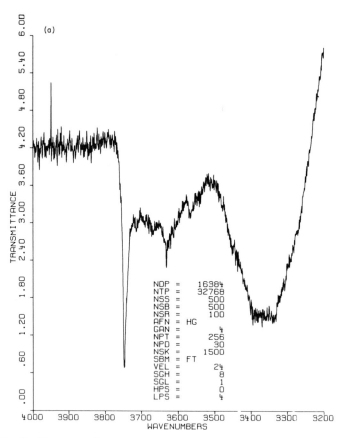

Fig. 3. Zeolite sample after vacuum activation at 500°C: An explanation of parameters in this and subsequent figures is given in Appendix A. (a) normal conditions; (b) 15-point smoothing.

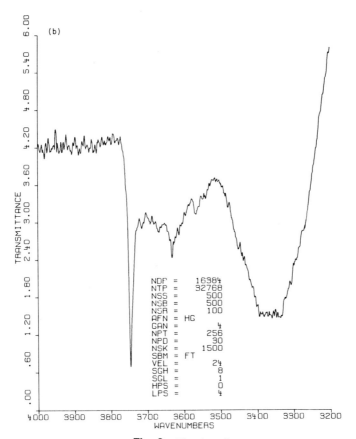

Fig. 3 *(Continued)*

obtained on a grating instrument does not show this band. As recommended by the staff at Nicolet, the problem was solved by omitting the "phase calculate" step in the sample interferogram calculation and using the phase calculated from the background interferogram in the "phase correct" step. This is referred to as using a "stored phase." This method works very satisfactorily and produces the true spectrum in every case (Fig. 4). For recently improved software, the use of a stored phase may not be necessary.

b. Changing Mirrow Velocity. Using a standard mirror velocity of 0.267 cm sec^{-1} and a room-temperature triglycine sulfate (TGS) detector, the limiting factor to the signal-to-noise (S/N) ratio of a spectrum is the speed of response of the detector. In the interferometer system used in the author's laboratory, it is possible to change the mirror velocity.

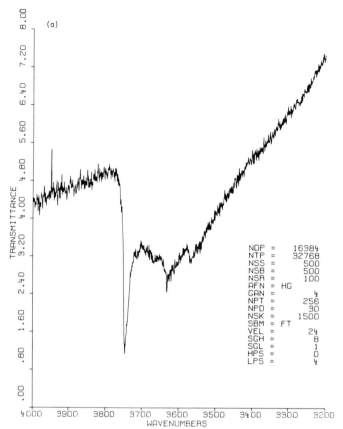

Fig. 4. Same zeolite sample as in Fig. 3: (a) using stored phase calculation; (b) stored phase and 15-point smoothing.

When the mirror velocity was changed to 0.133 cm sec^{-1} (VEL = 14, see Appendix A), the interferogram of the sample increased (correspondingly, the gain had to be decreased), and the resulting spectrum showed a much improved S/N ratio. We therefore improved the quality of the spectrum while increasing the time of data collection (keeping the number of scans constant).

    c. Cooled Detectors. There is, of course, another way to remedy the limitation imposed by the detector response—the use of a cooled detector. By using a liquid-nitrogen-cooled mercury–cadmium telluride (MCT) detector, we could increase the mirror velocity to 1.163 cm sec^{-1} and still obtain a perhaps tenfold increase in the S/N ratio of the spectrum (Fig. 5).

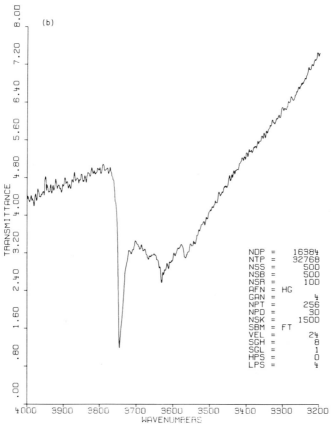

Fig. 4   (*Continued*)

While the use of a fast, cooled detector does provide great improvement, the rather high price of such a detector may present a problem.

None of the methods discussed so far has dealt with the actual problem of the transmittance spectrum in which some regions are transparent while others are opaque. There are, however, three further methods to deal with this.

d.   Electronic Filtering.   Commercial interferometers are equipped with electronic filters (Butterworth) of both the high-pass and low-pass types, which are used to remove the unwanted frequencies from the two ends of the spectrum. (For several settings of these filters, see Figs. 6 and 7.) Normally, they are set at values that allow presentation of the

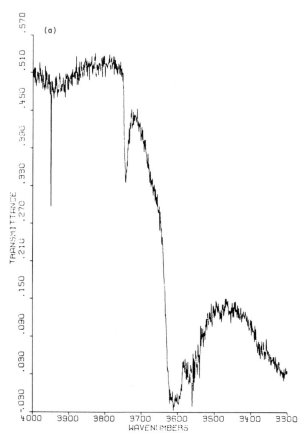

Fig. 5. Zeolite sample after vacuum activation at 350°C for 1000 scans and stored phase: (a) using TGS detector (VEL = 14); (b) using cooled MCT detector (VEL = 45).

spectrum in the 4000–400 cm⁻¹ region (at a velocity of 0.27 cm sec⁻¹, HPS = 1 corresponds to a 3-dB point of 100 Hz and LPS = 4 corresponds to a 3-dB point of 5 kHz). In the case of our problem samples, we are actually trying to remove the high-transmittance portion of the spectrum. By setting the high-pass filter to higher values, this can be achieved (see Fig. 7). Spectra of zeolites obtained with HPS = 3 (3-dB point of 2000 Hz) and with higher gains show much improved performances in the hydroxyl stretching region (Fig. 8). Unfortunately, these electronic filters do not provide sharp cutoffs, and a considerable amount of energy is lost in the region of interest.

e. Optical Filtering. Optical filters, however, can be obtained that provide very sharp cutoffs. We used a high-pass filter with a sharp cutoff

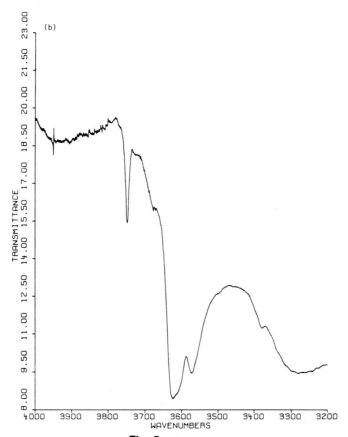

Fig. 5 (*Continued*)

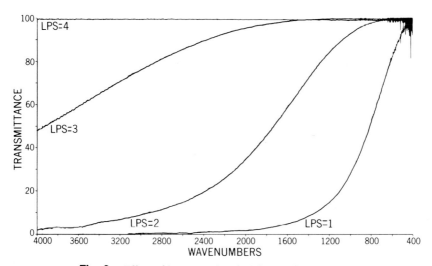

Fig. 6. Effect of low-pass electrical filters with HPS = 0.

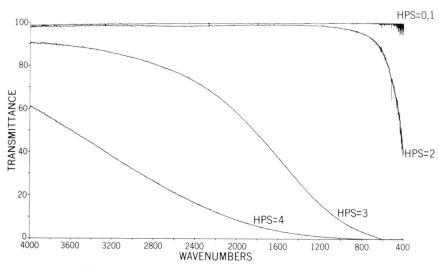

Fig. 7.   Effect of high-pass electrical filters with LPS = 4.

at about 2800 cm^{-1} with a 90% transmittance above this value (see Fig. 9). Using the filter in the sample beam, the total energy transmitted decreased greatly, and the gain could be increased. With the increased gain and the same number of scans, a much better hydroxyl region spectrum could be obtained (Fig. 10). Of course, it is quite clear that the same filter has to be used for obtaining the background spectrum also.

It seems hardly necessary to point out that while this method improves one part of the spectrum, the other part of the spectrum is lost. When both parts of the spectrum are needed, the only solution is to run the sample twice, once under normal conditions to get the spectrum of the high transmittance region and the second time using the described techniques to get a good spectrum of the hydroxyl stretching region.

Throughout the preceding discussion, we were concerned with samples where the low-transmittance region is at higher frequencies and the high-transmittance region is at lower frequencies. It is obvious that these methods can also be used for samples where the opposite is true. A low-pass electronic filter setting and a low-pass optical filter can be judiciously selected to achieve the same results as just described.

f. Smoothing.   Commercial interferometer systems have available a smoothing function that can be applied to the plotting of spectra. This facility is not confined to interferometric systems since dispersive infrared systems equipped with a computer also have a smoothing function available. While this gives the spectrum a more pleasing appearance (compare Figs. 4a and b), it does not, of course, change the actual data collected

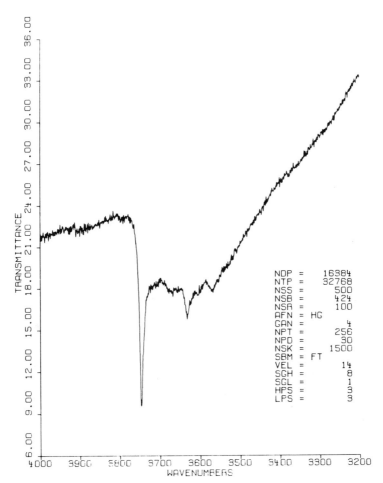

Fig. 8. Same zeolite sample in stored phase as in Fig. 3 with electrical filters at HPS = 3 and LPS = 3 with VEL = 14.

during the scans. However, we feel strongly that there is a possible danger in consistently using smoothing because it can lead to a sort of self-deception about the quality of spectra.

B. Infrared Spectra of Adsorbed Molecules

Förster and Schuldt (1977) have shown that when deuterium, nitrogen, and oxygen are adsorbed in NaA or NaCaA zeolites,† they can give rise

----

† See Appendix B for an explanation of zeolite nomenclature and structure.

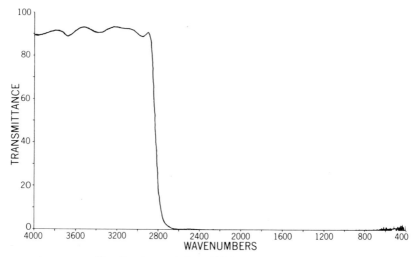

Fig. 9. Transmission of high-pass optical filter.

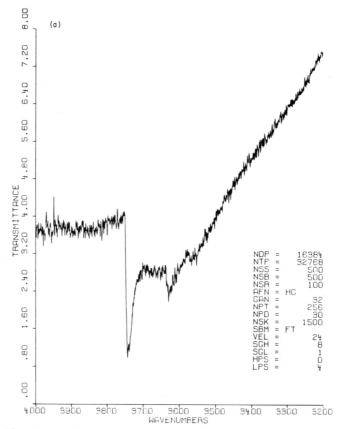

NDP	=	16384
NTP	=	32768
NSS	=	500
NSB	=	500
NSR	=	100
AFN	= HG	
GAN	=	32
NPT	=	256
NPD	=	30
NSK	=	1500
SBM	= FT	
VEL	=	24
SGH	=	8
SGL	=	1
HPS	=	0
LPS	=	4

Fig. 10. Same zeolite sample in stored phase as in Fig. 3, showing the effect of an optical filter: (a) VEL = 24 (normal conditions; (b) VEL = 14.

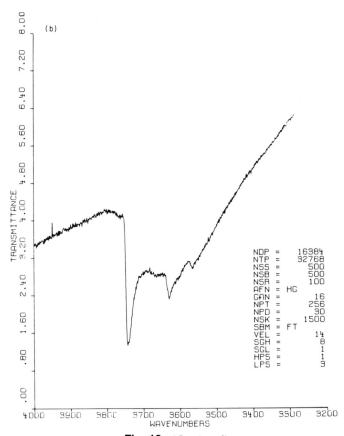

Fig. 10 (*Continued*)

to infrared bands. While these homonuclear diatomic molecules do not have infrared activity in the gas phase, in appropriate low-symmetry environments dipole moments can be induced, and thus infrared absorption can be observed. The favored sorption sites lie on a C_4 axis in NaA zeolite and on a C_3 axis in NaCaA zeolite. The energy levels of an A_2 rotor in an axial field have been calculated before. The most important feature is the lifting of the $J = 1$ degeneracy into $m = 0$ and $m = \pm 1$ levels. Only transitions between symmetric and antisymmetric states are allowed, as for the lowest levels a $J = 0 \rightarrow J = 0$ transition in case of o-D_2 and o-N_2 and $J = 1 \rightarrow J = 1$ transitions between $m = 0, +1$ levels for p-D_2, p-N_2, and oxygen. The observed infrared bands (see Fig. 11) are very close to the gas-phase fundamentals (from Raman spectra); the large red shift of the deuterium bands is caused by the small mass, and the blue shift of nitro-

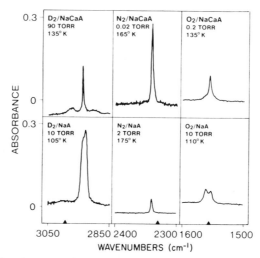

Fig. 11. Infrared spectra of deuterium, nitrogen, and oxygen inside zeolites NaCaA and NaA. Gas-phase fundamentals marked with a triangle. [From Förster and Schuldt (1977).]

gen is probably caused by the smaller quadrupole moment in the excited vibrational state. In another paper Förster and Schuldt (1978) also reported the observation of an infrared absorption band for hydrogen adsorbed on NaA and NaCaA zeolites.

Förster and Seeleman (1978) have observed the infrared spectra of *n*-butenes adsorbed on NaA and NaCaA zeolites of different degrees of Ca exchange. Because of their very strong lattice vibrational bands, zeolites have practically zero transmittance below 1200 cm^{-1}. They also have some weaker bands above 1200 cm^{-1}. In order to get rid of these, the "butene adsorbed on zeolite" spectra were ratioed against a background of the "zeolite only" spectrum, so that the spectrum of only the adsorbed species was obtained. All the bands of the butenes show the following features on adsorption: (i) small shifts in the band positions with respect to the gas phase; (ii) changes in band intensities; and (iii) a decrease in bandwidths. From the latter, the rotational behavior of the adsorbed molecule can be inferred. For example, in the case of *cis*-2-butene, the molecule has C_{2v} symmetry with the principal axis running through the center of the double bond. In this case, the band of the C=C stretching mode would have a change in dipole in the direction of the axis of rotation and would not be modulated by rotational motion. Therefore, this band would not be a good indicator of rotational quenching. On the other hand, the band of the CH$_3$ symmetric deformation vibration at 1410 cm^{-1}, with a dipole change perpendicular to the axis of rotation, should be sensitive to the quenching of rotational motion. Indeed, in the adsorbed state of *cis*-2-butene, this band becomes narrow, indicating that the rotational freedom of the adsorbed butene molecules is restricted by a high potential barrier to rotation.

Upon adsorption of each of the *n*-butenes on NaA or LiA zeolites, a sharp band is found in the C=C stretching region. On adsorption on a NaCaA zeolite, however, two bands appear in this region. It is assumed that in the latter case, the butene molecules are adsorbed on two different sites; i.e., on Na^+ ions and on Ca^{2+} ions. Because of their double charge, the calcium ions interact more strongly with the adsorbed molecules, resulting in a more pronounced spectral shift, than in the case of sodium ions. This assumption is well confirmed by the fact that on a CaA zeolite all the butenes give rise to only one C=C stretching band, the one characteristic of interaction with calcium ions.

Dalmon and co-workers (1975), in a series of articles, investigated the adsorption and reaction of CO on silica-supported nickel and on silica-supported nickel–copper alloys. The expected spectral range for the vibration of CO adsorbed on such systems was between 2200 and 1800 cm^{-1}. This region, however, contains a broad band of silica support, which makes the observation and measurement of small bands due to the adsorbed CO very difficult. Therefore, the pure spectrum of the adsorbed CO was directly computed from the spectra of the support recorded before and after adsorption.

It has been reported by several authors that CO adsorbed on nickel surfaces gives rise to two bands: the first at ~2060 cm^{-1} has been ascribed to a linear monodentate structure Ni–CO, while the second at ~1935 cm^{-1} has been assigned to a bridged structure

$$
\begin{array}{c}
O \\
\parallel \\
C \\
\diagup \quad \diagdown \\
Ni \qquad Ni
\end{array}
$$

The frequency of the CO in the complex is explained by a model first proposed by Blyholder (1964), which assumes a concerted electron transfer of the 3σ lone pair from carbon to metal and a back-donation of metal electrons to the lowest unfilled orbitals of CO (π^* antibonding). The larger the back-donation, the smaller the CO bond strength and the lower the CO stretching frequency. In this way it is possible to use the observed CO frequency as a measure of the electron-donating power of the metal.

One example of this model is given by Primet *et al.* (1977), who showed that in partially reduced nickel samples the frequency of adsorbed CO is at ~2075 cm^{-1}, while in fully reduced samples this frequency occurs at

~2045 cm^{-1}. In the system Ni–Ni^{2+} (partially reduced samples), nickel is expected to donate electrons to the Ni^{2+} ions. Thus, according to our scheme for CO chemisorption, the back-donation of nickel electrons to the π^* CO orbitals is decreased, resulting in an increase of the CO frequency. This agrees with the experimental observations since, for the incompletely reduced samples, the band is shifted toward higher frequencies.

Another example is presented by Dalmon *et al.* (1975) in their work on supported Ni–Cu alloys. As the copper content of the alloy increases, the frequency of both CO bands decreases (see Fig. 12). As copper has its d orbitals filled, it is expected to donate electrons to nickel atoms, with a filling of the d band; this situation increases the back-donation into the π^* orbital with consequent lowering of the CO stretching frequency. Indeed, the higher-frequency band decreases from 2060 to 2005 cm^{-1} in going from pure Ni to a Ni–93% Cu alloy.

As can be observed in Fig. 12, the intensity of the band due to the bridged species decreases with increasing copper concentration. This can be explained by a statistical consideration of the surface: as the concen-

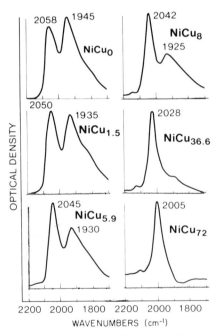

Fig. 12.   Infrared spectra of carbon monoxide irreversibly adsorbed at 25°C on Ni–Cu alloys. [From Dalmon *et al.* (1975).]

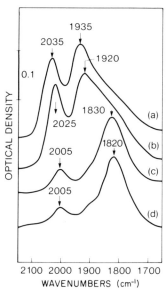

Fig. 13. Infrared spectra obtained during heat treatment of chemisorbed CO. (a) 15 ml of CO per gram of Ni adsorbed at 300°K, then heat treatment in closed system; (b) for 1 hr at 380°K; (c) for 2 hr at 480°K; (d) for 1 hr at 600°K. [From Martin *et al.* (1978).]

tration of copper atoms increases, the probability of finding two adjacent nickel atoms decreases. The authors, however, point out that the bridged species' concentration seems to drop off faster than the probability would indicate.

All the results discussed so far on the CO on supported nickel were obtained by room-temperature adsorption of the CO. Martin *et al.* (1978) have found that when CO was adsorbed on nickel at room temperature and then heated at 600°K with cryopumping of evolved gases, both the 2040 and 1935 cm^{-1} CO bands were completely removed and the trapped gas was CO_2, equal to half the volume of the initial CO. These results indicate that the reaction was the disproportionation of CO into carbon and gaseous CO_2: $2CO_{ads} \rightarrow CO_{2gas} + C$. Magnetic measurements suggest that the carbon probably formed a surface carbide of composition Ni_3C.

When the same experiment was carried out by heating at 600°K in a closed system, two new infrared bands appeared: a weak band at 2005 cm^{-1} and a strong band at 1820 cm^{-1} (Fig. 13). This indicates that the sites giving rise to the 2040 and 1935 cm^{-1} CO bands are no longer available.

When a sample that was heated with CO at 600°K with cryopumping (no infrared bands) was heated with CO_2 at 600°K, the same two bands,

2005 and 1820 cm^{-1}, appeared. Since these bands had been assigned to vibrations of adsorbed CO (CO$_{ads}$), it was deduced that the reverse reaction of the disproportionation was occurring:

$$CO_{2gas} + C \rightarrow 2CO_{ads}$$

The authors present a discussion about the nature of the surface species giving rise to the strong 1830 cm^{-1} band. While they are unable to exactly define this species, the evidence, based on infrared and magnetic measurements, strongly indicates a CO molecule attached to four nickel atoms: Ni$_4$–CO$_{ads}$. Combining a number of other experiments, the authors summarize the mechanism of the reaction that occurs when CO is heated with a nickel catalyst as

$$Ni-CO_{ads}, \; Ni_2-CO_{ads} \rightarrow Ni_4-CO_{ads} \rightleftarrows Ni_3C_{surface} + Ni-O_{ads} \rightleftarrows \tfrac{1}{2} CO_{2gas} + Ni_3C_{surface}$$
$$\Updownarrow$$
$$C_{interstitial}$$

In a study very similar to the case of Ni–Cu alloys mentioned previously, Primet and associates (1976) used infrared spectroscopy to examine the effect of the addition of silver to silica-supported palladium samples by observing the effect of silver on the type of chemisorbed carbon monoxide and the intensities and frequencies of the CO bands. They found that while using pure palladium the band (~1970 cm^{-1}) that predominated was usually attributed to a bridging complex involving CO; the addition of Ag atoms to Pd strongly favored the formation of the linear complex of CO as evidenced by the measured band at ~2050 cm^{-1} (Fig. 14). They also observed a shift in the position of this latter band with increasing amounts of Ag. According to the donation–back-donation model normally used to explain the chemisorption of carbon monoxide on metals, electron donation to the adsorbing metal should shift the CO vibration to lower frequencies as a result of increased π back-donation. The authors showed that the CO frequency of the adsorbed species decreased monotonically (from 2078 to 2049 cm^{-1}) with increasing silver content (from 0 to 65%). They therefore concluded that it was an electronic effect due to the changing silver content that was mostly responsible for the observed shift of the adsorbed CO band.

Chenery and Sheppard (1978), in their article on the practical performance of a FT infrared interferometer, also gave several good examples of the application of FT-IR spectroscopy to catalytic systems. In the spectrum of platinum supported on silica, three bands of the silica support at 2000, 1880, and 1640 cm^{-1} were clearly seen. When CO was adsorbed on this system and the spectrum ratioed against atmospheric background, it could be observed that the adsorbed CO gave a band at about 1850 cm^{-1}. The exact position and the intensity of this band would be very difficult

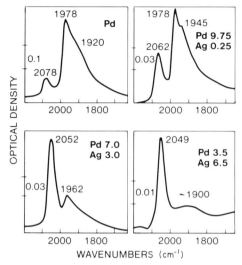

Fig. 14. Infrared spectra of carbon monoxide irreversibly adsorbed at 25°C on SiO$_2$-supported Pd–Ag alloys. [From Primet *et al.* (1976).]

to evaluate because of its overlap with the silica band. However, when the spectrum of the silica-supported platinum with CO is ratioed against the spectrum of the same sample before addition of CO, the bands of the silica support are cancelled, leaving only the two bands of adsorbed carbon monoxide (Fig. 15). Measurement of the exact position and the intensity of both bands now becomes an easy matter.

Okawa and co-workers (1977) reported that when ammonia was heated at 400°C with a magnesia-supported iron catalyst, it decomposed. The resulting spectrum showed two absorption bands, a weak one at 2200 cm^{-1} and a medium one at 2050 cm^{-1}. On heating the catalyst to 400°C in a vacuum, the intensity of the 2200 cm^{-1} band increased while the 2050 cm^{-1} band decreased. Both of these bands were attributed to nitrogen nondissociatively adsorbed on iron, with the 2200 cm^{-1} band attributed to an end-on linear form, and the 2050 cm^{-1} to a

$$\overset{\displaystyle N\equiv N}{\underset{\displaystyle Fe\qquad\quad Fe}{\diagup\qquad\diagdown}}$$

configuration. The experimental results indicate that the bridging form, upon heating, is converted into the linear form. It was also observed that the addition of ammonia removed the 2200 cm^{-1} band, i.e., replaced the nitrogen in this form, but did not affect the 2050 cm^{-1} band.

Metal loading of zeolites is usually accomplished by ion-exchanging

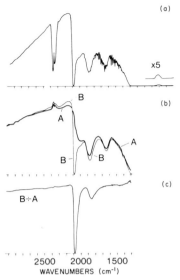

Fig. 15.   Infrared spectra of carbon monoxide adsorbed on SiO_2-supported platinum.
(a): single-beam spectrum from 620 scans; (b) ratios of A, SiO_2-supported platinum; and B,
CO adsorbed on SiO_2-supported platinum against atmospheric background; (c) ratio of
sample with CO against sample before addition of CO, based on 1000 scans. [From Chenery
and Sheppard (1978).]

the desired metal in the cationic form into the zeolite with subsequent reduction to the metal with hydrogen at a high temperature. Gallezot and
co-workers (1977) described a new process for metal loading, which involved the use of metal carbonyls. They used $Mo(CO)_6$, $Re_2(CO)_{10}$, and
$Ru_3(CO)_{12}$, which they deposited from the vapor phase on the HY zeolite
as adsorbent. They found infrared spectroscopy a most convenient tool
for the observation of what happened to such loaded zeolite samples on
gradual heating. In the case of $Mo(CO)_6$, heating at 70°C resulted in
the appearance of a new band at 920 cm^{-1}, which could be assigned to a
Mo–O bond, indicating that the molybdenum was attached to a framework oxygen. Heating above 150°C resulted in three CO stretching bands
at 2070, 2020, and 1985 cm^{-1}, which is consistent with a $Mo(CO)_4$ complex where the Mo atom is directly bonded to framework oxygen atoms.
In the case of $Re_2(CO)_{10}$, heating below 300°C led to similar results: a
band at 910 cm^{-1} representing a Re–O bond and CO bands characteristic
of a $>\!Re(CO)_3$ complex. Above 300°C, both types of bands disappeared,
and it was shown by CO adsorption that the rhenium was completely converted to very small metal particles encaged in the zeolite. Similarly, in
the case of $Ru_3(CO)_{12}$, heating of the metal-carbonyl-loaded zeolite ulti-

mately led to finely divided metal. The authors pointed out a significant advantage of their method: while ion exchange procedures introduce cations everywhere, especially on hidden sites that are unavailable for catalytic purposes, the metal-carbonyl complexes are too bulky to enter the sodalite cage and, therefore, the whole loading occurs in the supercages.

Bardet and associates (1978) studied the adsorption of carbon dioxide on nickel supported on silica, alumina, and magnesia. Heating the solid samples at 200°C in the presence of CO_2 in each case led to bands characteristic of CO adsorbed on nickel (2020 and 1830 cm^{-1}), indicating that the CO_2 molecule dissociated on the metal surface. In the case of the Ni on silica sample, these were the only bands observed. On the other hand, in the case of Ni on Al_2O_3 and Ni on MgO, bands appeared in the 1800–800 cm^{-1} region, which could be assigned to carbonate and bicarbonate groups attached to the surface. These bands represent the interaction of the CO_2 with the support rather than with the metal surface.

Candy and coworkers (1978) reported hydrogen adsorption on platinum supported on MgO or on a Y zeolite. At 300°K, the adsorption was reversible and led to two bands in the infrared spectrum at 2120 and 2060 cm^{-1}; both of these were assigned to Pt–H stretching vibrations. On very small particles of metal, e.g., 10 Å in the case of Pt on zeolite, only the 2120 cm^{-1} band appeared, while on larger Pt particles, e.g., 60 Å in the case of Pt on MgO, the 2120 and 2060 cm^{-1} bands were of about equal size. It was also shown that the species responsible for the 2120 cm^{-1} band was less strongly bonded than the 2060 cm^{-1} species. These results indicate that on larger Pt particles there are two different sites where the hydrogen can adsorb.

When hydrogen adsorption was carried out at 800°K, a strong band developed at 950 cm^{-1} which could not be removed on evacuation at 800°K. The authors assigned this band to the vibration of a hydrogen bridging between two or more Pt atoms.

The effect of metal-particle size on the properties of chemisorbed materials was further studied by Primet and associates (1975). They used nitric oxide (NO) as the adsorbate and platinum supported on alumina as the adsorbent. They prepared a series of samples with various sized platinum particles; the size distributions were characterized by electron microscopy and were found to be in the range of 15 to 35 Å. Nitric oxide was found to adsorb irreversibly on Pt and to give rise to a band between 1820 and 1775 cm^{-1}. The frequency of this band increased as the particle size of the metal decreased (Fig. 16). The idea that this variation in frequency was due to a difference in coverage of the metallic surface by NO was rejected because the optical density of the NO band plotted against metallic surface area gave a straight line, indicating a constant coverage of the

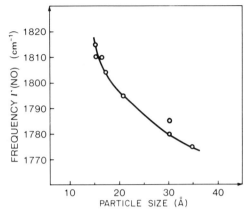

Fig. 16. Variation of the frequency of the ν(NO) band versus particle size, which was deduced from electron microscope measurements. Reprinted with permission Primet *et al.* (1975), *J. Am. Chem. Soc.* **95**, 3655. Copyright 1975 American Chemical Society.

metal by NO. However, the frequency variation was explained by an electronic effect. When NO bonds to a metal surface, the type of coordination is dependent on the relative energies of the π^* orbital of NO and the d orbitals of the metal; when the π^* orbitals are above the metal d orbitals, then donation of an electron from this orbital to the metal occurs, resulting in a NO^+ type molecule. On the other hand, when the π^* orbitals are below the metal d orbitals, then donation of an electron from the metal to the NO occurs resulting in a NO^--type molecule with a frequency below that of the neutral molecule. This latter effect is observed when NO is adsorbed on platinum. According to the donation–back-donation scheme previously discussed, the actual frequency of the adsorbed molecule (NO) depends on the electron-donating ability of the metal. The results presented by the authors can best be explained by stating that the smaller the particle size, the smaller the number of electrons available for back-donation and the higher the NO frequency. An interesting aspect of this work is the suggestion that the frequency of the adsorbed NO band can be used as a simple measure of the particle size of platinum–metal dispersions.

Another interesting application of the same principle was described by Primet and Sheppard (1976). They found that when CO was adsorbed on a bare nickel surface, the higher frequency band appeared at 2040 cm^{-1}, but when CO was adsorbed on a hydrogen-covered nickel surface, this band shifted to 2070 cm^{-1}. This result suggests that the adsorption of hydrogen changes the electron-donating property of the nickel surface. It can be pointed out that this leads to two significant conclusions. The first is that the presence of the bands at 2070 and 2040 cm^{-1}, due to the adsorption of CO on nickel, can be used to show whether hydrogen is present or not on the metal surface or to assess intermediate situations. The other is related to the fact that it was shown that high electron-donating properties of coadsorbed molecules led to low-frequency CO bands. The high frequency (2070 cm^{-1}) of the CO band in the presence of hydrogen suggests that the hydrogen is electron-attracting. This implies that the polarity of the metal–hydrogen bond is of the type $M^{\delta+}$—$H^{\delta-}$.

The first of these conclusions was applied by the authors to study the type of adsorption of ethylene on nickel. They observed that on a hydrogen-free surface with CO, the adsorption of ethylene gave rise to a CO band near 2070 cm^{-1}, which was attributed to a hydrogen-covered surface. They therefore concluded that the adsorption of ethylene was of the dissociative type, which occurred by the breaking of C–H bonds to form M–H bonds.

Sheppard and co-workers (1976), in a review on the chemisorption of hydrocarbons by metals, discussed some methods to distinguish among surface species that give overlapping absorption bands. They demonstrated the temperature-variation method using as an example the adsorption of ethylene on silica-supported platinum. They obtained spectra in the CH stretching region at temperatures ranging from -78 to 150°C. Two bands near 2885 and 2800 cm^{-1} always seemed to vary in intensity proportionately; these were assigned to the associatively adsorbed species MCH_2CH_2M. A third band at about 2930 cm^{-1} increased in relative intensity at higher temperatures; this band was assigned to the dissociatively adsorbed species M_2CHCHM_2. Another method was the coadsorption of CO. This molecule is so strongly adsorbed that it will frequently displace the more weakly adsorbed hydrocarbon species but will leave the more strongly adsorbed ones. In addition, the adsorption of CO will not give bands to interfere in the CH stretching region. As an example, the displacement-by-CO technique was applied to the adsorption of ethylene on silica-supported rhodium. The CO displaced two bands at 2887 and 2800 cm^{-1} (which were previously assigned to the associatively adsorbed MCH_2CH_2M species), but left three other bands between 2900 and 3000 cm^{-1} unchanged (Fig. 17). The authors compared the intensities of the CO bands after the displacement reaction with those of the CO bands on a bare metal surface and used this to estimate the fraction of the surface covered by the labile hydrocarbon species. It appeared that about 30% of the rhodium surface was covered by the MCH_2CH_2M species, although much more than 30% of the total intensity of the bands in the CH region was removed by CO.

There are large numbers of papers in the literature dealing with infrared

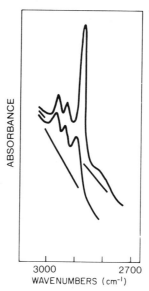

Fig. 17. Infrared spectrum of ethylene chemisorbed on rhodium (upper curve) and after partial displacement by carbon monoxide (lower curve). [From Sheppard *et al.* (1976).]

studies of the adsorption of ethylene on metal surfaces. From these it is concluded that ethylene adsorbs either dissociatively (giving the species M_2CHCHM_2) or associatively (giving MCH_2CH_2M), depending on the nature of the metal and on the conditions of the experiment. Prentice and associates (1976) reported evidence for a new form of ethylene chemisorbed on palladium and platinum. In addition to bands due to the previously mentioned species, they observed a band at ~ 1500 cm^{-1} and one at ~ 3020 cm^{-1}. The band at 1500 cm^{-1} was assigned to the C$=$C stretching vibration of a π-bonded ethylenic species, while the 3020 cm^{-1} band was in a position characteristic of olefinic CH stretching vibrations. The authors attributed their success in locating this long postulated species to the unusual capabilities of interferometer systems: exceptionally high sensitivity and the facility for accurate ratioing of spectra.

The assignment of the infrared bands of the associatively adsorbed ethylene species MCH_2CH_2M runs into some difficulties. Two bands in the CH region were assigned to this species: the strong band at 2885 cm^{-1} and the weaker one at 2800 cm^{-1}. The latter band was assigned to an overtone, leaving only the 2885 cm^{-1} band for assignment to a CH stretching fundamental. However, if the surface species MCH_2CH_2M has a planar skeleton, then the overall symmetry of the hydrocarbon group is C_{2v}, and then three out of the four CH stretching fundamentals should be infrared active. A way out of this dilemma was suggested by Pearce and Sheppard (1976). They resorted to a "metal surface selection rule," which states that only molecular vibrations that give dipole changes perpendicular to the metal surface will absorb radiation strongly. The picture from which this rule was derived is simple: when a dipole is placed over the surface of a conductor, it produces a virtual image below the surface (Fig. 18). When a dipole parallel to the surface vibrates, the "image" dipole changes are in the opposite direction, exactly cancelling any changes during the vibration. However, in the case of vibrations of a perpendicular dipole, the "image" dipole changes are in the same direction as the original, reinforcing the changes. Since for interaction with infrared radiation a dipole moment change of the whole system is required, it is easy to understand why only the perpendicular dipole vibrations give rise to infrared bands. In the case of the MCH_2CH_2M surface species (Fig. 19), only the A_1 vibration gives rise to a dipole change, which is perpendicular to the surface; the B_1 and B_2 vibrations produce dipole changes parallel to the surface; and there is no dipole change produced by the A_2 vibration. This model therefore explains why this species gives rise to only a single absorption band at 2885 cm^{-1}. What is more, it also specifies which of the molecular vibrations is responsible for this band.

Pearce and Sheppard (1976) reviewed the spectra of other molecules adsorbed on metals and found that the "metal surface selection rule" could be applied successfully in the interpretation (fuller understanding) of these spectra.

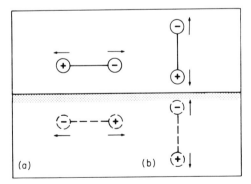

Fig. 18. The changes during the vibration of a dipole (a) parallel and (b) perpendicular to the surface of the metal. [From Pearce and Sheppard (1976).]

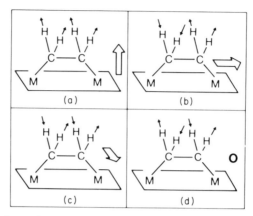

Fig. 19. The form of the C–H stretching normal modes of vibration of a MCH₂CH₂M surface species of symmetry C_{2v} which can result from the chemisorption of ethylene on a metal surface: (a) A_1 vibration; (b) B_1 vibration; (c) B_2 vibration; (d) A_2 vibration. [From Pearce and Sheppard (1976).]

Carbon monoxide adsorption on rhodium has been studied by many authors. Two recent articles by Primet and co-workers (Primet *et al.*, 1978; Primet, 1978) demonstrate well the complexity of such a system. The authors incorporated rhodium in the form of Rh(III) ions into the zeolite by ion-exchange techniques. When CO was adsorbed on such a sample at room temperature, it gave rise to four bands in the 2200–2000 cm⁻¹ region. By comparison with the infrared spectra of various Rh carbonyl compounds, these bands were assigned as follows: the two bands at 2172 and 2138 cm⁻¹ were assigned to CO adsorbed on Rh(III) species (The appearance of two CO bands was explained by either the formation of a gem dicarbonyl complex Rh(III) (CO)₂ or by two monocarbonyl complexes with the Rh(III) ions in different environments, the latter being favored by the authors.); the other two bands at 2115 and 2050 cm⁻¹ were assigned to two CO molecules adsorbed on Rh(I) ions. It was demonstrated that these latter two bands were enhanced by the addition of small amounts of water.

The presence of Rh(I) ions requires the reduction of Rh(III) ions by carbon monoxide. Such reduction reactions in solutions have been previously reported. The authors therefore suggest the following steps for the reaction:

fast reaction: Rh(III) + CO → Rh(III) – CO

reduction process: Rh(III) – CO + 2CO + H₂O → Rh(I)(CO)₂ + CO₂ + 2H⁺

In most of the preceding examples, emphasis was placed on the ability to remove the background spectrum that was caused by the support. The problem of interference from gas-phase spectrum did not arise, since all the examples dealt with chemisorption, where high adsorption can be achieved at very low pressures, but there are many cases where physical adsorption is investigated and the presence of significant gas pressure is necessary. In these cases the gas-phase spectrum overlaps the spectrum of the adsorbed species, and, unfortunately, for most of the gases usually studied—CO, CO_2, ethylene, acetylene—the spectrum of the gas is very strong. Even when special efforts are made to keep the optical pathlength in the gas as short as possible (see Section II), the gas-phase spectrum often seriously interferes with the bands of the adsorbed species. An example is given in Fig. 20. Spectrum (a) is that of a rhodium dispersion on silica in the presence of 200 Torr of CO. It was taken on a Digilab 14 FT-

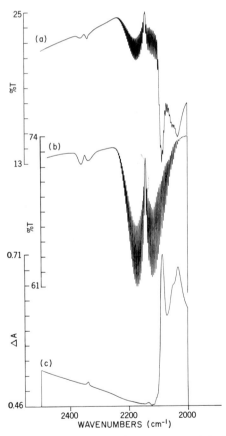

Fig. 20. Infrared spectra of carbon monoxide on supported rhodium: (a) sample in the presence of 200 Torr of CO; (b) gas phase CO only; (c) difference spectrum. [Mantz (1975).]

IR system (Mantz, 1975) in the short-pathlength cell devised by the author. Two bands due to adsorbed CO can be seen, at ~2085 and 2033 cm^{-1}, but their exact position and intensity cannot be determined because of the presence of the rotational lines of the gaseous CO. Spectrum (b) is that of gaseous CO determined by removing the sample from the beam. This way one can ensure that the spectrum of the gas is measured under exactly the same conditions (same cell, same pathlength, same pressure) as that of the sample. Spectrum (c) is the result of subtracting spectrum (b) from spectrum (a) (in absorbance). The two bands at 2085 and 2032 cm^{-1} are now clearly defined. In addition, features become apparent that could not be observed in spectrum (a): the shoulder at 2048 cm^{-1} and the small band at 2135 cm^{-1}. Also, a comparison of spectra (a) and (b) in the CO_2 region indicates that there seems to be a band at ~2337 cm^{-1} (the CO_2 was apparently present in the CO gas used), but it is only after the subtraction of the gas-phase CO_2 spectrum that the exact position and intensity of this band could be determined.

C. Far-Infrared Spectra

There has been very little work reported on catalysts in the far-infrared region. Butler and co-workers (1977) measured the spectra of a number of mono- and divalent cation-exchanged X and Y zeolites in the region 400–50 cm^{-1}. Some modifications of the usual sampling techniques were necessary for this region of the spectrum: the zeolites were examined either as a Nujol mull between polyethylene windows or as a powder dried on a silicon wafer from an aqueous slurry and then placed in a vacuum cell equipped with polyethylene windows. Some bands in this region were assigned to the vibrations of the cations against the zeolite framework. In most cases it was possible to go even one step further and assign certain bands as interionic vibrations of cations at specific sites in the zeolites, by taking into account the occupancy factors of the sites, the potential energy differences between the sites, and the behavior of the bands in the presence of adsorbed gases. These assignments were also supported by the fact that there is a nearly linear relation between the frequencies of these vibrations in each zeolite and $m^{-1/2}$ for each cation.

These authors also reported the effect of various adsorbates (water, THF, DMSO, pyridine, and methylene chloride) on the cation-vibration bands. For example, adsorption of water always decreases the frequency of these bands. This can be explained as the result of two effects: upon adsorption of polar solvents, the framework charge of the zeolites is delocalized; this charge delocalization reduces the force constants and vibrational frequencies of the cations. The cations may also be partially solvated, resulting in a further reduction of the vibrational frequency. The two processes, solvation and charge delocalization, operate together so that a steady decrease in the frequency assigned to cations in site II can be observed with increasing degrees of hydration. Interaction with adsorbed polar organic molecules results in shifts of the cation bands in the same direction but to a smaller extent than observed with water. Since the large cavities of the zeolites can hold only about 6 of the organic molecules, as opposed to over 30 water molecules, the delocalization of the framework charge will not be as effective as in hydrated zeolites, and direct cation–adsorbate interaction will be decreased for the cations in the large cages.

D. Reaction-Rate Studies

One important characteristic of interferometric systems has so far not been discussed: that is the extreme rapidity with which an infrared spectrum can be obtained with such a system. On all commercial rapid scan interferometers, a complete and identifiable infrared spectrum can be obtained in approximately 1 sec. It is, therefore, very surprising that FT-IR has been applied to reaction-rate studies only so rarely. While FT-IR-GC (which also depends on the rapidity of measurement) has been widely used, there is only one article in the literature that shows FT-IR applied to rate studies in a catalytic system.

Ceckiewicz and co-workers (1978) reported on the kinetics of transformation of 1-butene on NaHY zeolites with differing degrees of Na^+ ion exchange. The most characteristic band of 1-butene is that at 1630 cm^{-1}, which is assigned to the vinyl group CH_2=CH—. When 1-butene is contacted with a zeolite sample at room temperature, the 1630 cm^{-1} band gradually decreases, while bands at 1650 cm^{-1} and 3015 cm^{-1}, characteristic of the vinylene group —CH=CH—, gradually increase. The authors found that these changes occurred more rapidly on NaY zeolite than on partially exchanged NaHY zeolites (Fig. 21). These spectral changes can be ascribed to the isomerization 1-butene → 2-butene (cis and trans). As the reaction proceeds, other bands also appear in the spectrum: they are mostly characteristic of saturated hydrocarbons, especially isopropyl and isobutyl groups. They indicate that in addition to isomerization 1-butene → 2-butene (cis and trans). As the reaction proceeds, other bands also appear in the spectrum: they are mostly characteristic of saturated hydrocarbons, especially isopropyl and isobutyl groups. They indicate that in addition to isomerization, other reactions such as cracking, disproportionation, and oligomerization also occur. The authors concluded that even at room temperature, all these reactions occurred simultaneously, although their relative rates depended strongly on the type of zeolite that was used.

As mentioned earlier, it is surprising that so little use has been made of

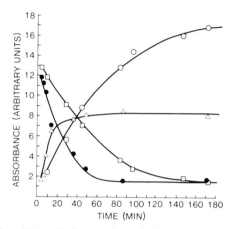

Fig. 21. Kinetics of ir bands characteristic of vinylene and vinyl groups during transformation of 1-butene on NaY-hydroxylated and on 23% exchanged NaHY zeolite samples: ●, $CH_2{=}CH{-}$(1630 cm^{-1}) on NaY; □, same on NaHY; △,$-CH{=}CH-$(3015 cm^{-1}) on NaY-hydroxylated; ○, same on NaHY. [From Ceckiewicz *et al.* (1978).]

FT-IR in reaction-rate studies, but it is our belief that it will not be very long before this technique will enjoy wider application to reaction-rate studies in catalytic systems.

IV. DIFFUSE REFLECTANCE SPECTROMETRY

Diffuse reflectance spectrometry has been widely used for the measurement of the ultraviolet–visible spectra of powders and turbid liquids. The most commonly used device for collecting diffusely reflected uv–visible radiation from a sample is an integrating sphere whose interior is coated with a nonabsorbing powder, such as MgO or BaSO$_4$. However, the use of an integrating sphere for mid-infrared spectroscopy presents some problems, mainly associated with the inefficiency of infrared detectors and the lack of suitable coatings for the interior of spheres with equivalent performance to that of MgO or BaSO$_4$ in the uv–visible region. A number of different types of mirror systems have been described to replace the integrating sphere; however, none of these has gained very wide application. Recently, Fuller and Griffiths (1978a,b) interfaced a fairly efficient ellipsoidal collecting mirror to a commercial rapid-scanning FT-IR spectrometer. They have demonstrated that diffuse reflectance infrared Fourier transform (which they named DRIFT) spectra of a variety of powdered samples can be measured rapidly at medium resolution and at high signal-to-noise ratio. They gave examples of diffuse reflectance spectra of pure organic compounds, of organic compounds dispersed in KBr, of sand, and of a variety of coals. To demonstrate the usefulness of this technique for microsampling, they showed good-quality spectra obtained on 180 ng of carbazole dispersed in KCl.

While the authors did mention the usefulness of DRIFT measurements to the study of adsorbed species on catalyst surfaces, they did not report any such experiments. The main difficulty in studying adsorbed molecules is that the sample has to be enclosed in an evacuable cell. However, we have enough confidence in the ingenuity of the Ohio University group to be assured that sooner or later they will design and build such a cell, which will be small enough to fit into the rather restricted sampling position of their apparatus.†

V. OTHER TECHNIQUES

There have been other techniques applying infrared spectroscopy to catalyst studies, especially to materials adsorbed on metal surfaces. However, these have been covered in recent reviews: internal and external reflection spectroscopy (Jakobsen, 1979) and reflection–absorption spectroscopy (Boerio 1978). Therefore, we did not discuss them in this chapter.

APPENDIX A: PARAMETERS FOR NICOLET 7199 FT-IR SYSTEM

NDP   number of data points collected per scan
NTP   number of transform points
NSS   number of scans taken in sample file
NSB   number of scans taken in background file
NSR   number of scans taken in reference file
AFN   apodization function (HG = Happ–Genzel)
GAN   gain of amplifier board
NPT   number of transform points used in phase calculation
NPD   number of data points used in phase calculation
NSK   number of skipped points between white light and the start of data collection
SBM   single beam, front (FT) or back (BK)
VEL   velocity of the moving mirror carriage
        $(14 = 0.133$ cm sec$^{-1})$
        $(24 = 0.267$ cm sec$^{-1})$
        $(45 = 1.163$ cm sec$^{-1})$
SGH   switched gain for beyond the 1024th point

† *Note added in proof:* This prediction became reality sooner than expected. Several such cells have been described (see talks given at the International Conference on Fourier Transform Infrared Specroscopy, Columbia, South Carolina, June 1981, and at the FACSS meeting, Philadelphia, Pennsylvania, September 1981).

SGL   switched gain for the first 1024 points
HPS   high-pass filter setting
                    3-dB roll-off (Hz)

0	10
1	100
2	250
3	1000
4	2000

LPS   low-pass filter setting
                    3-dB roll-off (kHz)

1	0.50
2	1
3	2
4	5

APPENDIX B: ZEOLITE STRUCTURES

Zeolites may be thought of as aluminosilicate framework structures enclosing cavities occupied by ions and water molecules, both of which have considerable freedom of movement, permitting ion exchange and reversible dehydration. The fundamental building unit for A, X, and Y zeolites is the sodalite unit [(a) in Fig. 22], the tetrahedral centers (silicon and aluminum atoms) being bridged by oxygen.

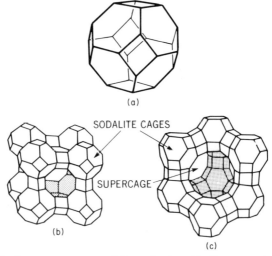

Fig. 22.   Zeolite structures; (a) sodalite unit; (b) cubically arranged sodalite units; (c) tetrahedrally arranged sodalite units.

Zeolite A can be thought of as cubically arranged sodalite units joined by square prisms [(b) in Fig. 22]. The nominal composition is $M_{12-2x}D_x(AlO_2)_{12}(SiO_2)_{12}$, where M is a monovalent and D a divalent cation. Zeolites X and Y can be thought of as tetrahedrally arranged sodalite units joined by hexagonal prisms [(c) in Fig. 22]. Their nominal compositions are

$$X = M_{86-2x}D_x(Al_2)_{86}(SiO_2)_{106}, \qquad Y = M_{56-2x}D_x(Al_2)_{56}(SiO_2)_{136}.$$

The inside of the sodalite unit is usually referred to as the "sodalite cage," while the large cavity between the sodalite units is sometimes referred to as the "supercage." There are cation positions in both types of cages, but adsorbed molecules can enter only the supercages.

When a zeolite contains sodium ions only, it is referred to as NaA, NaX, or NaY. Similarly, calcium-ion-containing zeolites are referred to as CaA, CaX, or CaY. The notation NaCaA refers to an A zeolite that contains both sodium and calcium ions.

ACKNOWLEDGMENTS

I wish to thank T. G. Kinisky for his help with much of the experimental work and Drs. C. S. Blackwell and P. H. Kasai for careful reading of the manuscript.

REFERENCES

Angell, C. L., and Schaffer, P. C. (1965). *J. Phys. Chem.* **69**, 3463–3470.

Bardet, R., Perrin, M., Primet, M., and Trambouze, Y. (1978). *J. Chim. Phys.* **75**, 1079–1083.

Bell, A. T. (1980). *In* "Vibrational Spectroscopies for Adsorbed Species" (A. T. Bell and M. L. Hair, eds.), ACS Symposium Series 137, pp. 13–35. Am. Chem. Soc.,Washington, D.C.

Blyholder, G. (1964). *J. Phys. Chem.* **68**, 2772.

Boerio, F. J. (1978). *In* "Applications of Polymer Spectroscopy," (E. G. Brame, Jr. ed.), pp. 171–184. Academic Press, New York.

Butler, W. M., Angell, C. L., McAllister, W., and Risen, W. M. (1977). *J. Phys. Chem.* **81**, 2061–2068.

Candy, J. P., Fouilloux, P., and Primet, M. (1978). *Surf. Sci.* **72**, 167–176.

Ceckiewicz, S., and Galuszka, J. (1976). *Reaction Kinet. Catal. Lett.* **5**, 257–263.

Ceckiewicz, S., Baranski, A., and Galuszka, J. (1978). *J. Chem. Soc., Faraday Trans. I* **74**, 2027–2036.

Chenery, D. H., and Sheppard, N. (1978). *Appl. Spectrosc.* **32**, 79–89.

Dalmon, J. A., Primet, M., Martin, G. A., and Imelik, B. (1975). *Surf. Sci.* **50**, 95–108.

Delgass, W. N., Haller, G. L., Kellerman, R., and Lunsford, J. H. (1979). "Spectroscopy in Heterogeneous Catalysis." Academic Press, New York.

Förster, H., and Schuldt, M. (1977). *J. Chem. Phys.* **66**, 5237.

Förster, H., and Schuldt, M. (1978). *J. Mol. Struct.* **47**, 339–343.

Förster, H., and Seeleman, R. (1978). *J. Chem. Soc., Faraday Trans. I* **74**, 1435–1443.

Fuller, M. P., and Griffiths, P. R. (1978a). *Am. Lab.* 69–80.

Fuller, M. P., and Griffiths, P. R. (1978b). *Anal. Chem.* **50,** 1906–1910.

Gallezot, P., Coudurier, G., Primet, M., and Imelik, B. (1977). "Molecular Sieves II," pp. 144–155. American Chemical Society, New York.

Hair, M. L. (1967). "Infrared Spectroscopy in Surface Chemistry." Dekker, New York.

Jakobsen, R. J. (1979). *In* "Fourier Transform Infrared Spectroscopy" (J. R. Ferraro and L. J. Basile, eds.), Vol. 2. Academic Press, New York.

Little, L. H. (1966). "Infrared Spectra of Adsorbed Species." Academic Press, New York.

Little, L. H. (1971). *In* "Chemisorption and Reactions on Metallic Films" (J. R. Anderson, ed.), Chapter 6, pp. 489–531. Academic Press, New York.

Mantz, A. W. (1975). Private communication.

Martin, G. A., Primet, M., and Dalmon, J. A. (1978). *J. Catal.* **53,** 321–330.

Okawa, T., Orishi, T., and Tamaru, K. (1977). *Z. Phys. Chem.* **107,** 239–243.

Pearce, H. A., and Sheppard, N. (1976). *Surf. Sci.* **59,** 205–217.

Prentice, J. D., Lesiunas, A., and Sheppard, N. (1976). *Chem. Commun.* 76–77.

Primet, M. (1978). *J. Chem. Soc., Faraday Trans. 1* **74,** 2570–2580.

Primet, M., and Sheppard, N. (1976). *J. Catal.* **41,** 258–270.

Primet, M., Bassett, J. M., Garbowski, E., and Mathieu, M. V. (1975). *J. Am. Chem. Soc.* **97,** 3655–3659.

Primet, M., Mathieu, M. V., and Sachtler, W. W. H. (1976). *J. Catal.* **44,** 324–327.

Primet, M., Dalmon, J. A., and Martin, G. A. (1977). *J. Catal.* **46,** 25–36.

Primet, M., Vedrine, J. C., and Naccache, C. (1978). *J. Mol. Catal.* **4,** 411–421.

Pritchard, J., and Catterick, T. (1976). *In* "Experimental Methods in Catalytic Research" (R. B. Anderson and P. T. Dawson, eds.), Vol. III, Chapter 7. Academic Press, New York. 281–318.

Sheppard, N., Chenery, D. H., Lesiunas, A., Prentice, J. D., Pearce, H. A., and Primet, M. (1976). *Proc. Eur. Congr. Mol. Spectrosc., 12th* pp. 345–353. Elsevier, Amsterdam.

Young, R. P., and Sheppard, N. (1967). *Trans. Faraday Soc.* **63,** 2291–2299.

2
ATOMIC EMISSION SPECTROCHEMICAL MEASUREMENTS WITH A FOURIER TRANSFORM SPECTROMETER

Gary Horlick

Department of Chemistry
University of Alberta
Edmonton, Alberta, Canada

R. H. Hall

Analytical Research
Syncrude Canada Ltd.
Edmonton, Alberta, Canada

W. K. Yuen

Saskatchewan Research Council
Saskatoon, Saskatchewan, Canada

I. INTRODUCTION

The development of effective simultaneous multielement analyses based on atomic emission spectrochemical methods is one of the major goals of a number of academic, government, and industrial laboratories.

FOURIER TRANSFORM
INFRARED SPECTROSCOPY, VOL. 3

A key aspect of this development is the design of spectrochemical measurement systems capable of simultaneously measuring spectral information over a wide range of wavelengths in the ultraviolet and visible spectral regions. To date, the dominant technique for the measurement of spectral information has been the dispersive system based on diffraction grating combined with photomultiplier tubes (PMT) or a photographic plate detector. The main types of instruments that result from the combination of these detectors with a grating dispersive system are the monochromator, the spectrograph, and the direct reading spectrometer, or polychromator. These instruments are shown schematically in Fig. 1.

In the monochromator–PMT combination (Fig. 1a) the spectral information is dispersed along a focal plane where an exit slit allows one resolution element at a time to reach the photomultiplier tube. The spectrum is obtained by rotating the dispersive element so that the spectral frequencies are sequentially detected by the PMT. This can be very slow if high resolution and/or wide spectral coverage is required. However, the photomultiplier is an excellent detector for the measurement of light intensity, having very good linearity, wide dynamic range, high sensitivity, and direct electronic output.

The second system, illustrated in Fig. 1b, is the spectrograph, in which a photographic plate is placed in the exit focal plane of the dispersive system. Although the photographic plate is capable of recording thousands of lines in a single exposure as a permanent record, it has a nonlinear response to light intensity, limited dynamic range, and tedious readout.

The direct reading spectrometer, or polychromator (see Fig. 1c) represents an attempt to retain the advantages of the PMT and also achieve simultaneous spectrochemical measurements. In this system an exit slit–PMT combination is mounted at each point in the exit focal plane of a dis-

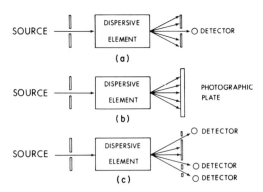

Fig. 1. Dispersive-based spectral information decoding systems: (a) monochromator; (b) spectrograph; (c) direct reader.

persive instrument at which measurements are to be made. Such systems provide simultaneous multiple-line detection and the number of lines detected is limited only by the physical problems incurred in fitting the detectors into the focal plane. Systems are available that provide the capability for detection of up to 60 emission lines simultaneously. However, although such systems provide reasonably powerful multichannel measurement capability, they, at best, detect only a very small fraction of the spectral information available in the exit focal plane.

Over the last few years much effort has gone into the development of spectrochemical measurement instrumentation to combine the best characteristics of all the systems shown in Fig. 1. *In particular, one would like a system combining the wide and detailed simultaneous wavelength coverage of the photographic plate with the near-ideal light-intensity measurement characteristics of the photomultiplier tube.*

A potential solution to this problem may be found through the use of modern image-sensor technology (Talmi, 1979). Such devices, which include silicon vidicons, secondary electron conduction vidicons, SIT vidicons, and silicon photodiode arrays, have all recently been applied to a variety of multichannel spectrochemical measurements. Although these sensors, when placed in the exit focal plane of a dispersive system, do provide continuous multichannel measurement capability, normally only a relatively narrow spectral window (50–5 nm) can be simultaneously observed and, in general, these sensors cannot match PMT performance in sensitivity and dynamic range.

Another potential approach to the overall problem is to dispense with the dispersive systems completely and to use a multiplex technique such as Fourier transform spectroscopy. In this technique the spectral information is encoded using a Michelson interferometer so that a single detector can be used to simultaneously measure a wide wavelength range (Fig. 2).

Fourier transform spectroscopy has been used extensively for spectral measurements in the infrared regions, but so far it has found little application for atomic spectrochemical measurements. Much of the impetus for

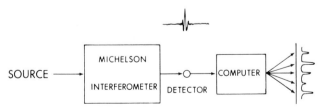

Fig. 2. Michelson-interferometer-based spectral information decoding system, for use in the multiplex technique of Fourier transform spectroscopy.

the development of infrared Fourier transform spectroscopy comes from the promise of achieving the multiplex (Fellgett's) advantage. In a simple analysis it can be shown that, in contrast to a scanning spectrometer, a Fourier transform spectrometer using a Michelson interferometer may achieve, in the same measurement time, a signal-to-noise ratio that is superior by a factor of $(M/2)^{1/2}$, where M is the number of resolution elements being observed (Marshall and Comisarow, 1975). It should be noted that not all authors agree on the exact magnitude of the multiplex advantage, with certain workers in the field indicating that it is a factor of two or more less than that just indicated. [See the discussions of Tai and Harwitt (1976), Harwitt and Tai (1977), Winefordner *et al.* (1976), Hirschfeld (1976), and Treffers (1977).] This controversy is not clearly resolved at this time. However, to whatever extent, the multiplex advantage remains a fundamental property of Fourier transform spectroscopy but only in measurement situations that are detector-noise limited. For atomic emission spectrochemical measurements, this is seldom the case. For such measurements signal-to-noise ratio considerations with respect to a Fourier transform spectrochemical measurement become very complex. The limiting noise in a system or a particular measurement may be associated with the detector, spectral background levels, source fluctuation, or signal level (where noise may be dependent on the square root of the signal level or dependent directly on the signal level or, indeed, any factor in between).

This whole area has not been well characterized in the literature except from the point of view of shot-noise limits, i.e., noise dependent on the square root of the signal level, see the discussions of Kahn (1959), Filler (1973), Chester *et al.* (1976), Winefordner *et al.* (1976), Hirschfeld (1976), Knacke (1978), and Luc and Gerstenkorn (1978), all of whom have reached approximately the same conclusions. Under shot- (photon-) noise-limited conditions, no signal-to-noise ratio advantage exists for the Fourier spectrometer except, perhaps, in the case of sparse narrow-line-width emission spectra. In addition, because of the distribution of the shot noise from a single strong line throughout the baseline of the complete spectrum, a so-called multiplex disadvantage for the measurement of weak lines in the presence of strong lines may result (Plankey *et al.*, 1974), but there can be an advantage if the lines are equally strong, However, it must be kept in mind that, in fact, most analytical atomic emission spectrochemical measurements using sources such as flames and inductively coupled plasmas are not shot-noise limited. At concentration levels above the detection limit, analyte flicker is often the limiting noise source. The effect of such noise in Fourier transform spectroscopy is difficult to quantitate as it is highly dependent on the noise-power spectrum of the flicker noise and the bandwidth of the Fourier transform spectrometer measurement system. We feel, at this time, that no definitive conclusions can be reached about the existence and importance of multiplex advantages and/or disadvantages for Fourier transform spectrochemical measurements of atomic emission sources without more experimental work. Such preliminary measurements will be presented later in this chapter. What the preceding overall discussion does mean is that the promise of a multiplex advantage, if any, cannot be a driving force in extending Fourier transform spectrometric techniques to atomic emission measurements.

Signal-to-noise ratio is not always the only and overriding consideration when carrying out a spectrochemical measurement (Horlick and Yuen, 1975). It is clear from the impressive success and capabilities of the Fourier transform technique in the infrared that some important advantages result from the nature of the instrumentation used to implement Fourier transform spectroscopy. These advantages are not dependent on the existence or realization of the multiplex advantage. Among them are the following: (1) spectra can be measured with a very accurate and precise wavenumber; (2) high resolution can be achieved in a relatively compact system; (3) the resolution function is easily controlled and manipulated as an inherent step in data reduction by the use of apodization techniques; and (4) computerization of the spectrometer is facilitated. In addition, with proper utilization of aliasing, a Fourier transform spectrometer can be very versatile in simultaneously covering a wide range of wavelengths. This is one aspect that can be a major limitation of array-based multichannel measurement systems such as vidicons and photodiode arrays as a consequence of the finite length and detector element density limitations of these devices.

Many of these capabilities and their importance for visible and ultraviolet Fourier transform spectroscopy were recognized by the French interferometry group several years ago. Excellent high-resolution atomic and molecular spectra obtained using Fourier transform spectroscopy in the visible and ultraviolet range have been reported (Gerstenkorn et al. 1977a,b) and the overall work of this group has been summarized by Luc and Gerstenkorn (1978).

In the next section the instrumentation developed in our laboratory for Fourier transform spectroscopy will be briefly described. Aliasing, a major consideration in making Fourier transform spectrochemical measurements in the ultraviolet–visible spectral region, will be considered in detail. Then, the application of Fourier transform spectroscopy to qualitative and quantitative atomic emission analysis will be discussed and illustrated.

II. ATOMIC EMISSION FOURIER TRANSFORM SPECTROMETER SYSTEM

A. Michelson Interferometer

A versatile Michelson interferometer system capable of Fourier transform spectrochemical measurements from the mid-infrared to the ultraviolet has been developed in our laboratories (Horlick and Yuen, 1978). This system will be briefly described, as all atomic emission spectrochemical measurements illustrated in this chapter were obtained with this system.

A block diagram of the Michelson interferometer system is shown in Fig. 3, and a schematic of the interferometer itself is shown in Fig. 4. The interferometer has three optical inputs: a He–Ne laser, a white light (tungsten bulb) source, and the spectral signal of interest. All optical signals, the He–Ne laser, the white light source, and the spectral signal of interest share the same beam splitter and Michelson mirrors. Laser fringe referencing is used in the interferometer system to sequence digitization and to control the velocity of the moving mirror using a phase-locked

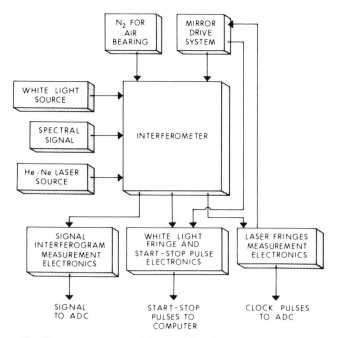

Fig. 3. Block diagram of the Michelson interferometer system.

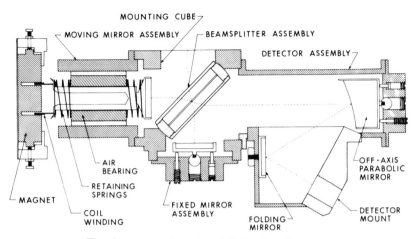

Fig. 4. Schematic of the Michelson interferometer.

loop. The mirror drive system consists of an electromechanically driven mirror supported by an air bearing. The mirror movement is repetitive, and both the scan rate and length can easily be set or altered. With the control signals derived from the white light interferogram and the laser fringes, repetitive signal interferograms can be precisely time averaged. This system uses a unique pretrigger approach that allows acquisition of any desired number of signal interferogram data points both before and after the zero difference position.

It is important to consider in more detail the fact that in this interferometer system all three optical signals share the same Michelson mirrors and beam splitter assembly. This is a major difference when compared to most current commercial systems, which use two or even three independent interferometers for these signals. We have found the single interferometer design to be both simple and effective. In particular, alignment and the maintenance of alignment are greatly facilitated as only one mirror needs to be adjusted.

The major consequence of the single-interferometer design is the coincidence in time (and mirror position) of the white light and spectral signal interferograms. The occurrence of the white light interferogram is normally used as a precise start flag to ensure coherent addition of repetitive interferograms. In the single-interferometer design that we have used, it is not effective simply to generate a start pulse from the white interferogram as it and the central fringe of the signal interferogram occur coincidentally. It is difficult to calculate acceptable spectra from "one-sided" interferograms with *no* points acquired before the central fringe of the signal interferogram because of phase correction problems (Horlick and Yuen, 1975; Yuen and Horlick, 1977). In fact, this is a major reason why, in most designs, a separate interferometer is used for the white light channel. If this is done, then one of the Michelson mirrors of the white light channel can be offset so that the white light central fringe occurs several hundred laser fringes before the spectral signal central fringe.

However, with a single-interferometer system, an alternative approach must be used to control time averaging. The basic philosophy of our approach is to use the white light interferogram to generate a precise *stop* pulse. In our system we start taking data at some imprecise point before the position of zero pathlength difference. When the white light interferogram occurs, it is used to start precisely a modulo–N counter that counts the laser digital clock. The counter overflow pulse is the precise stop pulse and is used to terminate data acquisition. Repetitive interferograms can be coherently time averaged by sequencing addition from the last point acquired rather than the first, in a sense time averaging backward. This is easy to achieve with software. This system has been described by Horlick and Yuen (1978), and this reference should be consulted for further details.

B. Detectors

Two basic types of detectors have been used for the atomic emission measurements, a silicon photodiode detector for near-ir measurements (600–1000 nm) and photomultiplier tubes for visible and ultraviolet measurements. A standard 1P28 photomultiplier has been used for most visible measurements (300–600 nm) and a Hamamatsu R166 solar blind photomultiplier tube for the ultraviolet measurements.

In order to make measurements with the interferometer system coupled to PMTs, it was necessary to use a modified dynode chain. Two modifications were necessary. First, the chain had to be optimized for the detection of ac signal changes by capacitively coupling the last three stages. Second, the gain of these latter stages had to be increased also.

Due to the multiplex nature of the measurement, the PMT must respond to a consistently high light level as all the intensity and frequency information is present in the optical signal all of the time. By the time this high level signal reaches the last couple of dynodes in the chain, the number of secondary-emission-produced electrons is enormous. A problem with such a large number of electrons exists in attempting to focus them onto the last dynodes. When operated as a linear potential difference dynode chain, a high number of electrons are simply lost from the last dynodes due to the dynode's inability to electrostatically focus the elec-

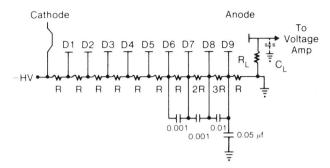

Fig. 5. Schematic of the modified photomultipler tube dynode chain.

trons. This results in a decay of signal quality even at moderate levels. A solution to the electrostatic focusing problem is discussed in the EMI literature.† By increasing the resistance value over the last two dynodes, a larger potential is allowed to develop, thereby increasing their electrostatic focusing capacity. The changes made are illustrated in Fig. 5. Discussions and implementations of similar changes are available in the literature by Lytle (1974) and Harris *et al.* (1976).

C. Computer System and Computations

The minicomputer data acquisition and processing system used with the Michelson interferometer consists of a PDP 11/10 processor with 32K core, an RK05 cartridge disk, a VT-11 graphics processor, a Zeta 130 incremental plotter, and a DEC laboratory peripheral system containing a 12-bit analog-to-digital converter and sample and hold input amplifier. All software was developed under the RT-11 operating system. Both a machine language FFT (8K) and a FORTRAN floating point FFT (4K) are available on the system.

It is important to note that we acquire and process full double-sided interferograms with our system, i.e., an equal number of points on both sides of the central fringe. All spectra shown in the next section were calculated from 4096-point double-sided (2048 each side) interferograms. With a double-sided interferogram we can calculate a spectrum without using any phase-correction procedures, i.e., by calculating the amplitude spectrum. This capability of our Fourier transform spectrometer has been crucial for our measurements of line emission spectra in the uv–visible

† EMI Photomultiplier Tube Catalogue, EMI Gencom Inc., 80 Express St., Plainview, New York 11803.

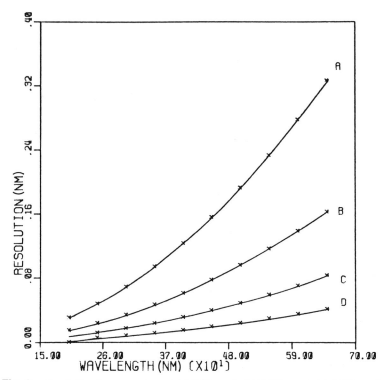

Fig. 6.   Resolution in nanometers at (a) 7.75 cm^{-1}; (b) 3.875 cm^{-1}; (c) 1.988 cm^{-1}; and (d) 0.994 cm^{-1}.

spectral region. The phase-correction procedures for a "one-sided" interferogram utilized in mid-ir Fourier transform spectroscopy are designed primarily for absorption spectra measured using broadband sources and tend to be inadequate with line emission spectra.

With this length of interferogram, the resolution was 7.75 cm^{-1} across the complete spectral range. While this might be considered moderate to low resolution in the infrared, it does provide a resolution of about 0.08 nm at 300 nm, which is reasonable for atomic spectroscopy in the ultraviolet (see curve (a) in Fig. 6). The uv–visible resolution in terms of nanometers that can be obtained with wavenumber resolutions of 3.875, 1.988, and 0.994 cm^{-1} are also shown in Fig. 6. Thus, with only moderate drive capability (compared to some ir interferometer systems), quite adequate resolution can be obtained for atomic measurement in the uv–visible region.

D. Atomic Emission Sources

Three types of atomic emission sources were used in conjunction with the interferometer system: hollow cathode lamps, flames, and an inductively coupled plasma. The hollow cathode lamps were those standardly used for atomic absorption measurements and were operated at conventional conditions.

1. Flames

Both air-C_2H_2 and N_2O-C_2H_2 flames were used. The flame system employed a Varian Techtron manual gas flow control box and a Varian Techtron nebulizer–burner support assembly, both from the model AA6 spectrophotometer. The normal slot burner heads were not used because of excess flicker present in the resulting flames. Custom cylindrical burner heads were made for both air-C_2H_2 and N_2O-C_2H_2 following the design of Aldous *et al.* (1970) and Winefordner and Haraguchi (1977). This system provides a very stable cylindrical flame that can be sheathed with nitrogen if so desired.

2. Inductively Coupled Plasma

A Plasma-Therm argon radio-frequency inductively coupled plasma (ICP) source was used for the ICP measurements. The ICP specifications and normal operating parameters are listed in Table 1. The optical link between the plasma source and interferometer proved to be nontrivial due to the combination of the collimated light input requirements of the interferometer and the high degree of spatial resolution required on the plasma

TABLE 1

Inductively Coupled Plasma
Operating Conditions

Parameter	Specification
Frequency	27.12 MHz
Power	1.5 kW
Coolant argon flow	18 min⁻¹
Nebulizer argon flow	0.8 min⁻¹
Observation height	22 mm above load coil
Observation aperture	5.0 mm
Solar blind PMT voltage	600 V
1P28 PMT voltage	400 V

plume. Various authors have shown the influence of spatial effects when working with a plasma-monochromator-based system [see, for example, Boumans and deBoer (1975)]). Small changes in the nebulizer gas flow rate, coolant gas flow rate, and/or observation window result in very large changes in the analyte signal level [refer to Edmonds and Horlick (1977) for a detailed study]. The following optical arrangement was found to provide adequate spatial selectivity for this preliminary work. Using a 5-cm diameter, 10.0-cm focal length quartz lens, an image of the plasma plume was projected on a 5-mm diameter aperture centered 22 mm above the load coil of the plasma torch assembly. A 12.5-cm focal length, 5-cm diameter quartz lens, placed 12.5 cm from the aperture and about 30.0 cm from the entrance aperture of the interferometer, provided the collimated light required. All three units, the plasma, lenses, and interferometer were optically aligned using a He–Ne laser.

III. ALIASING

One of the major points to consider in utilizing Fourier transform spectroscopy in the ultraviolet and visible regions is the aliasing question. With the standard He–Ne reference laser, the basic sampling interval for the interferometer system is 0.6328 μm, which is one cycle of the laser line modulation. This means that the shortest wavelength of light that can be properly sampled without aliasing is 1.266 μm (7901 cm^{-1}). Clearly then, to work in the ultraviolet–visible region, either the sampling rate must be increased or aliasing must be tolerated.

Aliasing (Horlick and Malmstadt, 1970; Malmstadt *et al.*, 1974) refers to the undersampling of modulation frequencies in the interferogram. Normally, in Fourier transform spectroscopy, aliasing is avoided as the undersampled modulation frequencies show up as spurious spectral information (fold-over). This can be very confusing when broadband spectral information is present. However, with line spectra, as commonly measured in atomic emission spectroscopy, it is possible to actually use aliasing to advantage (Horlick and Yuen, 1975; Yuen and Horlick, 1977).

As previously stated, the basic sampling interval of a He–Ne laser-referenced interferometer is 0.6328 μm. While 1.266 μm is the shortest wavelength that can be sampled without aliasing, shorter wavelengths can be accurately measured via aliasing. In a more general sense, the 0.6320-μm sampling interval allows a bandwidth of 7901 cm^{-1} of spectral information to be uniquely sampled. These bandwidths occur at fixed locations and are summarized in Table 2, which will now be illustrated through the use of several examples.

A practical illustration of the use of aliasing occurs in the measurement

TABLE 2

Aliasing Table for Various Sampling Rates

Region number	÷4 Sampling rate (2.5312 μm)[a] Region (cm⁻¹)	Region (μm)	÷2 Sampling rate (1.2656 μm)[a] Region (cm⁻¹)	Region (μm)	Direct sampling rate (0.6328 μm)[a] Region (cm⁻¹)	Region (μm)	×2 Sampling rate (0.3164 μm)[a] Region (cm⁻¹)	Region (μm)	×4 Sampling rate (0.1582 μm)[a] Region (cm⁻¹)	Region (μm)
1	0–1975.35	∞–5.0624	0–3950.7	∞–2.5312	0–7901.4	∞–1.2656	0–15,802.8	∞–0.6328	0–31,605.6	∞–0.3164
2	3950.70–1975.35	2.5312–5.0624								
3	3950.70–5926.05	2.5312–1.6875	7901.4–3950.7	1.2656–2.5312						
4	7901.40–5926.03	1.2656–1.6875								
5	7901.40–9876.75	1.2656–1.0125	7901.4–11,852.1	1.2656–0.8437	15,802.8–7901.4	0.6328–1.2656				
6	11,852.10–9876.73	0.8437–1.0125								
7	11,852.10–13,827.45	0.8437–0.7232	15,802.8–11,852.1	0.6328–0.8437						
8	15,802.80–13,827.45	0.6328–0.7232								
9	15,802.80–17,778.15	0.6328–0.5625	15,802.8–19,753.5	0.6328–0.5062	15,802.8–23,704.2	0.6328–0.4219	31,605.6–15,802.8	0.3164–0.6328		
10	19,753.50–17,778.15	0.5062–0.5625								
11	19,753.50–21,728.85	0.5062–0.4602	23,704.2–19,753.5	0.4219–0.5062						
12	23,704.20–21,728.85	0.4219–0.4602								
13	23,704.20–23,679.55	0.4219–0.3894	23,704.2–27,654.9	0.4219–0.3616	31,605.6–23,704.2	0.3164–0.4219				
14	27,654.90–25,679.55	0.3616–0.3894								
15	27,654.90–29,630.25	0.3616–0.3375	31,605.6–27,654.9	0.3164–0.3616						
16	31,605.60–29,630.25	0.3164–0.3375								
17	31,605.60–33,580.95	0.3164–0.2978	31,605.6–35,556.3	0.3164–0.2812	31,605.6–39,507.0	0.3164–0.2531	31,605.6–47,408.4	0.3164–0.2109	63,211.2–31,605.6	0.1582–0.3164
18	35,556.30–33,580.95	0.2812–0.2978								
19	35,556.30–37,531.65	0.2812–0.2664	39,507.0–35,556.3	0.2531–0.2812						
20	39,507.00–37,531.65	0.2531–0.2664								
21	39,507.00–41,482.35	0.2531–0.2411	39,507.0–43,457.7	0.2531–0.2301	47,408.4–39,507.0	0.2109–0.2531				
22	43,437.70–41,482.35	0.2301–0.2411								
23	43,457.70–45,433.05	0.2301–0.2201	47,408.4–43,457.7	0.2109–0.2301						
24	47,408.40–45,433.05	0.2109–0.2201								
25	47,408.40–49,383.75	0.2109–0.2025	47,408.4–51,359.1	0.2109–0.1947	47,408.4–55,309.8	0.2109–0.1808	63,211.2–47,408.4	0.1582–0.2109		
26	51,359.10–49,383.75	0.1947–0.2025								
27	51,359.10–53,334.45	0.1947–0.1875	55,309.8–51,359.1	0.1808–0.1947						
28	55,309.80–53,334.45	0.1808–0.1875								

[a] Sampling interval.

TABLE 3

Major Emission Lines of Na, Li, K, Rb, Cs in the
Near-Infrared

Element	Emission line (nm)	Emission line (cm⁻¹)	Region number (in Table 2)
Na	589.00	16,978	9
	589.59	16,961	9
Li	670.78	14,908	8
K	766.49	13,046	7
	769.90	12,989	7
Rb	780.02	12,820	7
	794.76	12,582	7
Cs	852.11	11,736	6
	894.35	11,181	6

of the flame emission spectra of lithium, potassium, rubidium, and cesium with a Fourier transform spectrometer. The major emission lines of these elements fall in the near-ir and the visible spectral regions and are listed in Table 3. Let us imagine, for a moment, that we can take samples at any multiple of the He–Ne laser sampling interval. In order to measure the flame emission spectrum of Li, K, Rb, and Cs without aliasing, it would actually be necessary to take samples at twice the normal rate of the He–Ne reference clock, i.e., at 0.3164-μm intervals. This would achieve an unaliased sampling bandwidth of $15,802$ cm⁻¹, sufficient for the measurement of these spectral lines. This is shown schematically in Fig. 7a. In this schematic spectrum, the amplitudes of the lines are arbitrary. Now, in reality, we actually sample at 0.6328-μm intervals covering a bandwidth of 7901 cm⁻¹. This results in fold-over of the spectral lines, as shown in Fig. 7b, the 7901–$15,802$ cm⁻¹ region being aliased into the 0–7901 cm⁻¹ region. However, in the case of this particular measurement, the overlap is not serious since there is no signal in the 0–7901 cm⁻¹ region. Therefore, as long as one is aware of aliasing, the frequency axis can be properly assigned and the spectrum interpreted.

Now, what if a spectral signal greater than $15,802$ cm⁻¹ (shorter than 632.8 nm) is present. An example, again from flame emission spectroscopy, in which sodium is present along with the previous four elements, is shown schematically in Fig. 8. An unaliased measurement of this spectrum would require sampling at a rate equivalent to four times that of the

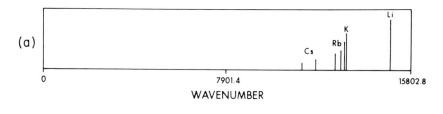

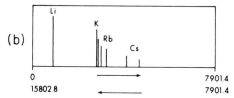

Fig. 7.   Schematic spectrum of Li, K, Rb, and Cs (a) and the aliased schematic spectrum for a 0.6328-μm sampling interval (b).

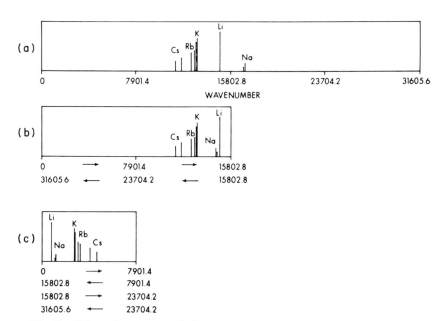

Fig. 8.   Schematic spectrum of Li, K, Rb, Cs, and Na (a) and the one-level (b) and two-level (c) aliasing schematic spectra.

standard He–Ne rate, resulting in a bandwidth of 31,605 cm^{-1}. This is shown in Fig. 8a. One level of aliasing (Fig. 8b) results in the aliasing of the Na doublet between the K and Li lines and two levels of aliasing to reach the basic He–Ne sampling rate results in fold-over as shown in Fig. 8c. In this final spectrum the Na doublet has been aliased twice and the Li, K, Rb, and Cs lines once; thus there is spectral information from more than one aliased region and interpretation is less trivial. However, as long as the aliased lines do not overlap exactly, interpretation is possible. In fact, aliasing could be considered an advantage as it compresses spectral information covering a wide bandwidth into a smaller bandwidth making more efficient use of a given number of data points.

The actual flame emission spectrum of these alkali metals (Na, Li, K, Rb, and Cs), as measured with our Fourier transform spectrometer, is shown in Fig. 9. An air–C_2H_2 flame was used as the excitation source, and a Si photodiode was the detector. Fifty scans were averaged, and the interferogram was apodized with a Gaussian function. The interferogram was sampled at the direct rate derived from the He–Ne laser. The digitized interferogram is shown in Fig. 10 where the middle 3000 points of the 4096-point double-sided interferogram are plotted.

Since the sampling interval was 0.6328 μm, all the spectral lines shown in Fig. 9 are aliased as previously explained. The wavenumber axis is labeled for points within regions 5 to 8 (see Table 2). The Li, K, Rb, and Cs peaks lie in this region, but the Na doublet is actually aliased in from region 9 to 12.

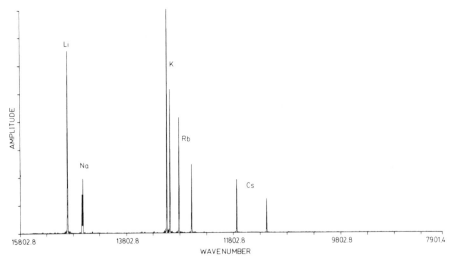

Fig. 9.   Flame emission spectrum of the alkali metals.

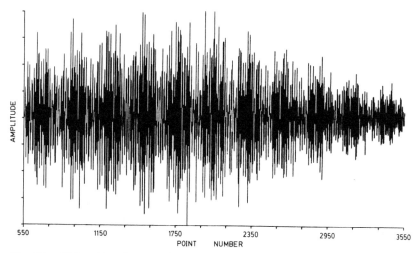

Fig. 10. Digitized interferogram for the spectrum of the alkali metals shown in Fig. 9.

A final example of aliasing is shown in Fig. 11 for the emission spectrum of Cr. The chromium triplets at 427 nm and 360 nm are shown as they would appear unaliased with a sampling bandwidth of 31,605 cm^{-1} (Fig. 11a). If the sampling bandwidth were 15,802 cm^{-1}, the triplets would be folded over into the 0–15,802 cm^{-1} region pivoting about the 15,802

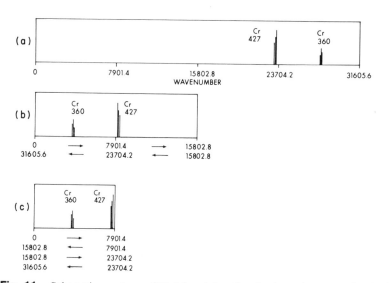

Fig. 11. Schematic spectrum of Cr (a) and the aliased schematic spectra for (b) the 15,802 cm^{-1} bandwith and (c) the 7901 cm^{-1} bandwidth.

cm^{-1} point (Fig. 11b). Finally, with the standard 7901 cm^{-1} bandwidth, the Cr 427-nm triplet is folded over again about the 23,704 cm^{-1} point (Fig. 11c). The actual chromium emission spectrum from a hollow cathode lamp source, as measured with our Fourier transform spectrometer, is shown in Fig. 12. Only about 3000 points of the 4096-point spectrum are plotted. The wavenumber axis is for points within regions 13 to 16 of Table 2. The Cr triplet at 360 nm is expanded in Fig. 12b and the Cr triplet

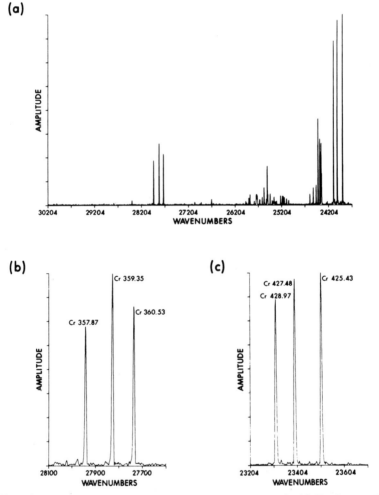

Fig. 12. Spectrum of a Cr hollow cathode lamp as measured with Fourier transform spectrometer (a), along with expanded plots of Cr triplets at (b) 360 and (c) 427 nm.

at 427 nm is expanded in Fig. 12c. This triplet (427) is aliased from regions 9 to 12 as just mentioned. The wavenumber axis for this expanded plot was relabled for points within this region.

These examples make it clear that although aliasing may be complex it is quite predictable and manageable for line emission spectra. The overlapping of spectral regions due to aliasing is extensively tabulated in Table 2, based on sampling rates derivable from the He–Ne laser 632.8-nm line. In sorting out aliased regions in a real spectrum, a knowledge of detector spectral response and bandwidth of optical components is invaluable. In addition, optical filters can be used to illuminate and/or identify aliased spectral information. However, one of the most effective methods of identifying aliased lines is to carry out preliminary measurements with the × 2 and × 4 sampling rates. The frequency of the reference laser modulation can be increased by the use of optical techniques, such as double passage of the laser through the interferometer, or by electronic frequency multiplication using phase-locked loops (Malmstadt *et al.*, 1974). We have found the latter method to be quite useful in our own laboratories. One might well ask why we do not sample routinely at the higher rates derived by the above methods. It is because the high sampling rate necessary to sample ultraviolet–visible modulation frequencies without aliasing quickly results in a prohibitively large number of data points that must be digitized in the interferogram to achieve reasonable resolution.

An illustration of the use of both frequency multiplication and then aliasing to clarify and then optimize measurement bandwidth is shown in Fig. 13. Using phase-lock-loop frequency multiplication techniques, the sampling rate of our interferometer was doubled to provide a sampling interval of 0.3164 μm. This provided sufficient bandwidth to sample the emission lines of Li, K, Rb, and Cs without aliasing. This was illustrated in Fig. 7. The actual flame emission spectrum resulting from such a measurement is shown in Fig. 13a. For this figure only, 512-point interferograms were processed. With this limited number of data points, the resolution was quite poor. The potassium doublet was not resolved at all and the rubidium doublet was only partially resolved. For the second spectrum (Fig. 13b), the interferogram was sampled at the direct rate of the He–Ne laser. The lines were aliased as depicted in Fig. 7, but the resolution was doubled due to the fact that the sampling interval was twice as long, i.e., with the same number of data points (512 in this case), the interferogram had been sampled to twice the optical retardation as was the case for the spectrum shown in Fig. 13a. If the sampling interval is lengthened to 1.2656 nm (a data point every other laser fringe), the bandwidth becomes 3950 cm^{-1}. This is the $\div$ 2 sampling rate, and the overlapping of spectral regions due to aliasing is tabulated in columns 4 and 5 of Table 2.

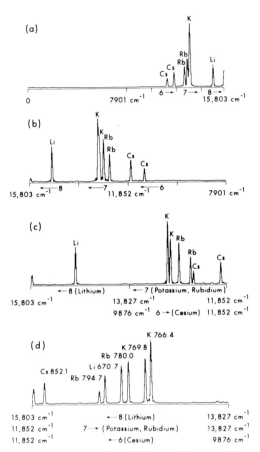

Fig. 13. Flame emission spectra of Li, K, Rb, and Cs using (a) × 2, (b) direct, (c) ÷ 2, and (d) ÷ 4 sampling rate.

The spectrum resulting from an interferogram sampled at this rate is shown in Fig. 13c. The resolution was again doubled, the potassium doublet was resolved, and the cesium lines (in region 6) aliased into the potassium–rubidium region (region 7). Finally, if the sampling is made even coarser (2.5312 μm), the bandwidth becomes 1975 cm^{-1}. The resulting spectrum is shown in Fig. 13d. Aliasing was extensive with region 8 (lithium), region 7 (potassium–rubidium), and region 6 (cesium) all overlapped. Resolution was again doubled and, in fact (with 512 points), was only a factor of 2 poorer than for the 4096-point data shown earlier in Fig. 9. From this series of spectra (Fig. 13) we see that aliasing can be used to advantage in optimizing spectral coverage and resolution when making atomic spectrochemical measurements with a Fourier transform spectrometer.

It is clear from the illustrations and discussion in this section that aliasing considerations are uniquely important to atomic emission spectrochemical measurements using a Fourier transform spectrometer. It is nearly impossible to eliminate aliasing in practical applications, thus the existance of aliasing must always be kept in mind when interpreting the spectra that will be shown in the next two sections.

IV. ATOMIC EMISSION MEASUREMENTS

A. Hollow Cathode Lamps

Hollow cathode lamps (HCL), such as those used in atomic absorption spectroscopy, are excellent atomic emission sources with which to test the basic performance capabilities of the interferometer. Using these lamps, we assessed the spectral range, resolution, aliasing complications, and wavelength accuracy under normal operating conditions. In addition, the multielement lamps provided an indication of the simultaneous multielement detection capability of the system.

The following hollow cathode lamps were used as atomic emission sources: 1. Si; 2. Mg; 3. Cu–Zn–Pb–Sn; and 4. Ca–Al–Mg. For lamps 1–3, a solar blind photomultiplier tube (R166, Hamamatsu T.V. Co. Ltd.), operated at a potential of 600 V, was used as a detector with a peak spectral response at 220 nm falling to 10% at 160 and 300 nm. For lamp 4, a standard 1P28 photomultiplier tube (RCA), operated at 400 V, was employed as a detector with a peak spectral response at 350 nm falling to 10% peak response at 185 and 650 nm. Both tubes employed the modified dynode chain outlined in Section II. The optical lamp–interferometer interface was accomplished by placing the cathode of the lamp at the focal point of a 6.6-in. focal length, 2-in. diameter, aluminum-coated, off-axis parabolic mirror. A preamplifier provided any amplification necessary for the detector signal before entering the data acquisition system. The frequency bandpass was set at 300 Hz and 10 kHz for lower and upper limits, respectively. Double-sided interferograms, 4096 points in length, were acquired and Gaussian apodization was employed. As part of the FFT processing, one level of zero filling was employed yielding 4096 unique spectral points. These are the standard conditions under which all of the following interferograms have processed unless otherwise stated. The only variation may lie in the number of signal-averaged interferograms present in the final interferogram. For the hollow cathode lamps, 50 signal-averaged interferograms were used to obtain the final interfero-

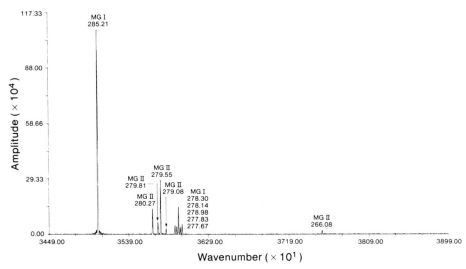

Fig. 14. Magnesium hollow cathode lamp emission spectrum.

gram. This amounted to a total observation time of 50 sec and a total experimental time of 2 min.

The spectra of the magnesium and silicon hollow cathode lamps, as measured with the solar blind photomultiplier tube, are shown in Figs. 14 and 15. All the lines of the Mg spectrum fall within one aliased region,

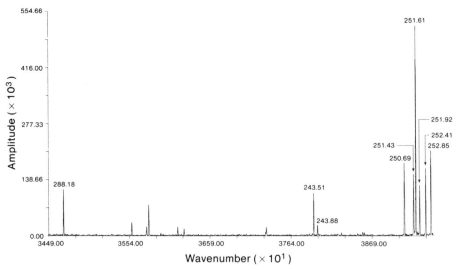

Fig. 15. Silicon hollow cathode lamp emission spectrum.

regions 17–20 of Table 2 (31,605–39,507 cm^{-1} or 316.4–253.1 nm). The resolution at 277.5 nm is 0.06 nm, which is quite adequate to resolve the neutral atom emission quintuplet at 277.8 nm.

In order to interpret the Si spectrum (Fig. 15), two aliased regions must be considered. The 288.1 nm line occurs in the same region as the Mg lines of Fig. 14 but all others are in the next aliased area, i.e., regions 21–24 of Table 2 (39,507–47,408 cm^{-1} or 253.1–210.9 nm). The observed and literature wavelengths for both of these spectra are tabulated in Table 4. Note the high degree of accuracy in the observed wavelengths. It is important to note that the signal channel and the laser channel must be very precisely aligned relative to each other if an accuracy in the range of one part in 10^7 is desired. Although no special alignment procedures were carried out, the single-interferometer design of our instrument facilitates simultaneous alignment of the signal and laser channels. All literature

TABLE 4

Si and Mg Hollow Cathode Lamp Emission Lines

| Element | Observed lines | | Literature wavelength[a] (nm) |
	Wavenumber (cm^{-1})	Wavelength (nm)	
MgI	35,059.4	285.23	285.21
MgII	35,777.3	280.29	280.27
	35,738.5	279.81	279.81
	35,769.2	279.57	279.55
	35,830.7	279.09	279.08
MgI	35,931.2	278.31	278.30
	39,950.9	278.16	278.14
	35,973.4	278.98	277.98
	35,991.9	277.84	277.83
	36,014.0	277.67	277.67
MgII	37,581.3	266.09	266.08
Si	39,542.9	252.89	252.85
	39,613.4	252.44	252.41
	39,690.4	251.95	251.92
	39,737.7	251.65	251.61
	39,767.8	251.46	251.43
	39,885.1	250.72	250.69
	41,012.2	243.83	243.88
	41,059.3	243.55	243.52
	34,700.5	288.18	288.16

[a] From Zaidel' *et al.* (1970).

wavelengths for *all tables* were taken from Zaidel' *et al.* (1970). The wavelength values in Zaidel' *et al.* (1970) are listed to three decimal places in nanometers. All values (experimental and literature) are reported and listed to two decimal places, as previously reported results from our laboratory illustrated the wavelength jitter of our instrument to be about 0.03 nm with 4096-point transform (Yuen and Horlick, 1977). The work of Luc and Gerstenkorn (1978) illustrates the very high degree of wavelength accuracy and precision obtainable with a higher-resolution interferometer.

The spectrum of the Cu–Zn–Pb–Sn multielement hollow cathode lamp provides information about the spectral response of the current system. The spectrum is shown in Fig. 16 and tabulated in Table 5. As with the Si spectrum, two aliased regions are overlapped. Notably absent are the 213.8-, 206.1-, and 202.5-nm lines of Zn along with the 220.4- and 217.0-nm lines of Pb. These lines are well within the spectral response of the solar blind PMT, and thus beam splitter efficiency and/or mirror reflectivities are probably limiting spectral response at these far-uv wavelengths.

Measurements with the 1P28 photomultiplier tube detector and the Ca–Al–Mg hollow cathode lamp illustrate a potential problem arising from multiple-aliased regions and also related to dynamic range limitations in Fourier transform spectroscopy. The spectral response of the 1P28 PMT extends from about 180 to 650 nm. In addition to lines of the specified elements, a hollow cathode lamp emits the spectrum of the filler

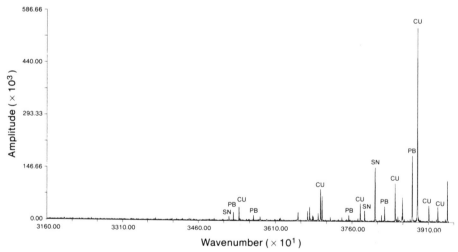

Fig. 16. Cu–Zn–Pb–Sn hollow cathode lamp emission spectrum.

TABLE 5

Cu–Zn–Pb–Sn Hollow Cathode Lamp Emission Lines

| | Observed lines | | Literature |
| | | | wavelength[a] |
Element	Wavenumber (cm^{-1})	Wavelength (nm)	(nm)
CuI	35,403.2	282.46	282.44
	38,188.3	261.86	261.84
	40,120.4	249.25	249.22
	40,950.0	244.20	244.16
PbI	35,297.0	283.31	283.33
	35,688.8	280.20	280.22
	37,548.8	266.32	266.32
	38,252.6	261.42	261.43
	38,800.3	257.73	257.79
	40,381.2	247.64	247.66
	—	—	220.40
	—	—	217.00
SnI	34,924.7	286.33	286.35
	35,211.3	384.00	284.03
	36,948.1	270.65	270.66
	38,886.3	257.16	257.13
	39,268.0	254.66	254.66
	40,267.4	248.34	248.37
Zn	—	—	213.80
	—	—	206.10
	—	—	202.50

[a] From Zaidel' *et al.* (1970).

gas, which is normally neon or argon. The Ca–Al–Mg lamp is filled with neon. The first spectrum obtained of this lamp is shown in Fig. 17. It is completely dominated by filler gas lines, as neon has several very strong lines in the 500–650 nm region. Thus this is an example of spectral information from one aliased region—in this case an undesirable spectral region—dominating the desired aliased region. One cannot simply turn up the gain to detect the less intense lines of interest because in Fourier transform spectroscopy all information must stay "on scale." However, what one can do is optically filter the spectral signal. An optical bandpass filter (Corning 7-51) was used to isolate the spectral region from about 310 to 420 nm. The resulting spectrum is shown in Fig. 18 and now the element lines of interest can be seen (see Table 6). Thus aliasing must, in some cases, be carefully controlled.

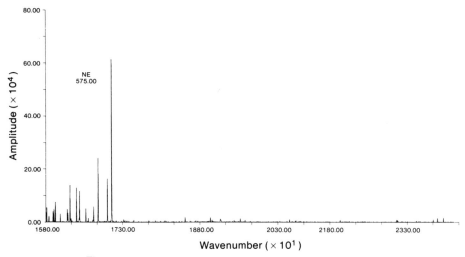

Fig. 17. Ca–Al–Mg hollow cathode lamp emission spectrum.

The use of hollow cathode lamps as atomic emission sources shows the simultaneous multiline detection capability of the interferometer system. The spectra are of high resolution and high wavelength axis accuracy, although aliasing complications do exist. The interferometer's dynamic range problem was illustrated and a solution to such a problem was of-

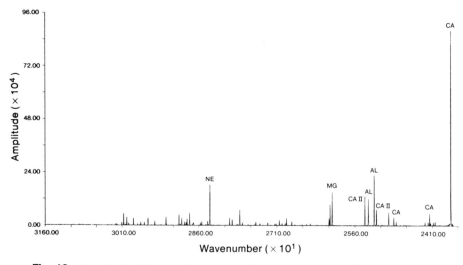

Fig. 18. Ca–Al–Mg hollow cathode lamp emission spectrum including filter no. 7-51.

TABLE 6

Ca–Al–Mg Hollow Cathode Lamp Emission Lines

	Observed lines		Literature
Element	Wavenumber (cm^{-1})	Wavelength (nm)	wavelength[a] (nm)
CaI	23,659.8	422.66	422.67
	23,347.2	428.32	428.30
	23,314.4	428.92	428.94
	23,243.1	430.24	430.25
	23,214.1	430.77	430.77
	23,156.2	431.85	431.87
	22,596.7	442.54	442.54
	22,548.4	443.49	443.50
	22,448.1	445.47	445.48
CaII	22,212.7	396.89	396.85
	21,990.8	393.43	393.37
AlI	25,240.1	396.20	396.15
	25,354.9	394.40	394.40
MgI	26,050.5	383.87	383.83
	26,091.0	383.27	383.23
	26.110.3	382.99	382.93
Ne	26,928.4	371.36	371.31
	27.067.3	369.45	369.42
	26,828.1	372.74	372.71
	27,289.2	366.45	366.41
	27,825.6	359.38	359.35
	28,402.1	352.09	352.05
	28,445.1	351.56	351.52
	28.794.3	347.29	347.26
	29,255.4	341.82	341.77
	29,672.2	337.02	336.99
	29,984.8	333.50	333.49

[a] From Zaidel' *et al.* (1970).

fered through optical filtering of the incoming signal. This may or may not be viable in all situations, depending on the proximity of the line of interest and the range-limiting emission line. There appears to be a lower wavelength system sensitivity of about 240 nm. This is most likely due to the germanium beam splitter no longer acting as a beam splitter at these low wavelengths. Depending on the emission wavelengths of interest, this may or may not be a severe limitation. Overall, the interferometer-based atomic emission system appears very promising for simultaneous multielement analysis. Let us now proceed to further studies using various atomic emission sources.

B. Flame Atomic Emission Spectroscopy

Since the rise of atomic absorption spectroscopy, flame atomic emission spectroscopy has not been widely used except for determinations of the alkali metals and perhaps the alkaline earths, although the method has been somewhat more widely used since the development of the nitrous oxide–acetylene flame (Pickett and Kortyohann, 1969). We have applied our Fourier transform spectrometer to both qualitative and quantitative measurements using air-C_2H_2 and N_2O–C_2H_2 flames. These measurements will be briefly discussed in this section.

The qualitative measurement capability of the Fourier transform spectrometer for the simultaneous measurement of the flame emission spectra of the alkalis (Na, Li, K, Rb, and Cs) has already been illustrated in Section III (see Fig. 9). These spectra were obtained using an air–C_2H_2 flame and a Si photodiode detector. Lithium was chosen to test the basic quantitative capability of the system. Solutions were run containing lithium concentrations ranging from 30 to 0.03 ppm. The spectra of three of these solutions in the immediate vicinity of the Li 670.8-nm line are shown in Fig. 19, and the complete analytical curve is shown in Fig. 20. The slope of the log–log plot is 1.12 ± 0.02 and the correlation coefficient is 0.998. The arrows on the plot in Fig. 20 indicate 95% confidence levels based on six

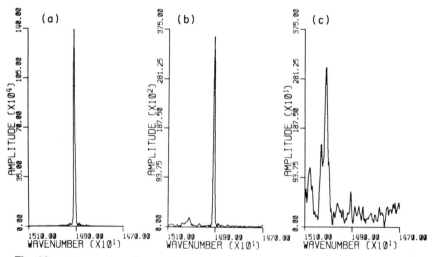

Fig. 19. Air–acetylene flame emission of (a) 30-ppm Li; (b) 1-ppm Li; (c) 0.03-ppm Li.

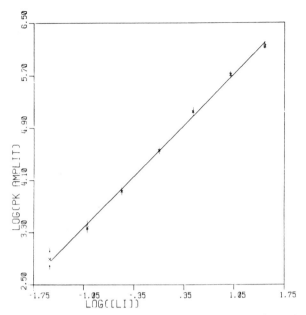

Fig. 20.   Analytical curve for Li flame emission data in Fig. 19 for 0.03 to 30 ppm.

repetitions completed at each concentration. The signal-to-noise ratio (relative standard deviation)$^{-1}$ for the 30-ppm lithium peak was 109 and that for the 0.03-ppm peak was 8.

In order to assess some of the noise properties of this Fourier-transform-based quantitative determination, plots were made of the standard deviation of the lithium peak and the average of the standard deviations for four baseline values as a function of lithium peak amplitude. These plots, shown in Fig. 21, indicate that the noise is localized on the peak and is not distributed into the base line. Curve (a) seems to have three distinct regions. The portion of the curve at low peak amplitude with a slope near zero indicates the presence of a constant-level background noise, probably originating in the detector of its associated electronics. The sloping portion of the curve indicates a steady increase in noise with emission intensity—a situation characteristic of a source fluctuation noise-limited system. At high lithium emission intensity, the curve again flattens, indicating the presence of a constant noise level independent of emission intensity. Photodiode saturation is not indicated since analytical curve response is still linear within this concentration range. This curve was reproducible over several experiments, and no detailed explanation could be provided for its behavior at high concentration values.

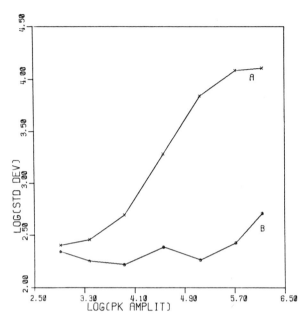

Fig. 21. Standard deviation versus Li peak amplitude; (a) Li peak standard deviation; (b) base line standard deviation.

A common analysis is that for Na and K levels in blood serum. Normal physiological blood serum levels for sodium range from 3000 to 4000 ppm and for potassium from 100 to 300 ppm. Using the interferometer system with the Si photodiode detector, these elements can be simultaneously determined using Li as an internal standard. A portion of the flame emission spectrum of a solution containing 20-ppm Li, 35-ppm Na, and 2-ppm K is shown in Fig. 22. These concentrations of Na and K simulate those that would result from a 100-fold dilution of a serum sample. The simultaneously determined analytical curves over the physiological range are shown in Fig. 23. Note that these curves are plots of analyte emission intensity divided by Li emission intensity (internal standard) versus analyte concentration. The interferometer-based measurement nicely provides the simultaneous multiline measurement capability necessary for this measurement.

For the practical measurement by flame emission spectroscopy of elements other than the alkalis, it is necessary to use photomultiplier tube detectors. Using a 1P28 photomultiplier tube detector and an air–C_2H_2 flame, Ca and Na have been determined. Partial emission spectra for Na at the 30-, 1-, and 0.01-ppm levels are shown in Fig. 24. The intensity distortion of the sodium doublet in Fig. 24a is due, primarily, to photomultiplier tube saturation effects encountered at high signal levels. Over this concentration range the slope of the analytical curve (log–log plot) was 0.9 ± 0.1 and the correlation coefficient was 0.99. The detection limit was 3 ppb. Noise distribution data for both Ca and Na were similar to that illustrated for Li, i.e., noise was localized on the signal peaks. We are not exactly sure of the source of the limiting noise in these flame emission measurements. It is probably some type of analyte flicker noise. Whatever the source, it does not appear to distribute noise throughout the base line as has been predicted for the so-called multiplex disadvantage.

Finally, we present some qualitative results obtained with the $N_2O-C_2H_2$ flame. Our particular burner system proved to be too noisy for reliable quantitative measurements. The emission spectrum in this flame of an aqueous solution containing 20 ppm each of Cr, Mn, and Cu is shown in Fig. 25. Note that the copper doublet does not appear in its normally observed intensity ratio. The use of an optical filter (Corning 0-52), used to eliminate the hydroxyl band emission at 306.4 nm, has a slight attenuating effect on the copper lines at 327.4 and 324.8 nm. Although not separately tabulated, wavelength accuracies are comparable to those illustrated for hollow cathode lamps in Section IV.A. Also note the measurement of the CrI lines at 520.82 and 520.55 nm via aliasing into the vicinity of the CrI 360-nm triplet.

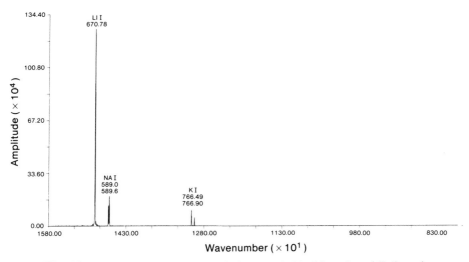

Fig. 22. Flame emission spectrum of Li (20 ppm), Na (35 ppm), and K (2 ppm).

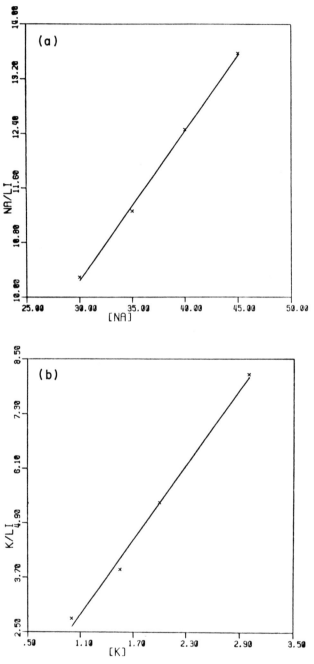

Fig. 23. Analytical curves for data in Fig. 22: (a) sodium, 30–45 ppm; (b) potassium, 1–3 ppm.

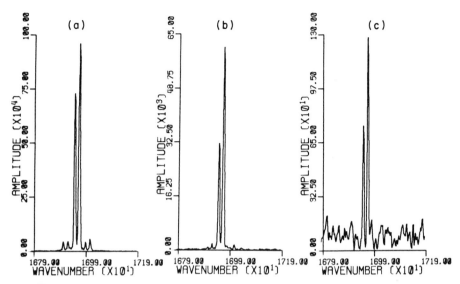

Fig. 24.   Air-acetylene flame emission spectrum of (a) 30-ppm Na; (b) 1-ppm Na; (c) 0.01-ppm Na.

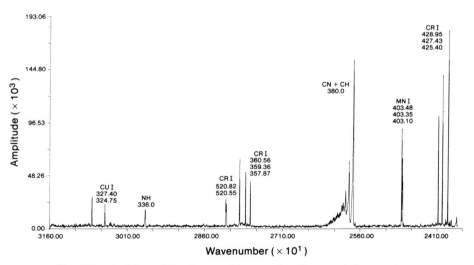

Fig. 25.   Cu, Mn, and Cr nitrous oxide–acetylene flame emission spectrum.

C. Inductively Coupled Plasma
   Optical Emission Spectroscopy

Inductively coupled plasma is currently the most effective source for atomic emission spectrochemical analysis. Numerous authors have discussed and reviewed the ICP and its applications (Dahlquist and Knoll, 1978; Winge *et al.*, 1977; Greenfield *et al.*, 1976). In most commercial configurations the ICP source is coupled to a direct reading spectrometer in order to facilitate simultaneous multielement analysis. However, direct reading spectrometers are inflexible as spectromchemical measurement systems. In this section we present some of our initial measurements carried out using the interferometer-based system for the measurement of ICP emission signals.

The ICP emission spectrum for a 1-ppm aqueous solution of Ca is shown in Fig. 26. Note that this relatively low concentration results in a high-quality, high signal-to-noise ratio spectrum. For the 1-ppm Ca solution, the 393.37-nm CaII line amplitude had a relative standard deviation of 0.6%. Also note that with this source the ion lines dominate the calcium emission spectrum, which is indicative of the high excitation energies available in this source.

The large amount of excitation energy provided by the plasma source not only affects the species of a particular element excited to emission but also the variety of elements capable of being excited. In contrast to the flames previously discussed, many elements not excited now readily un-

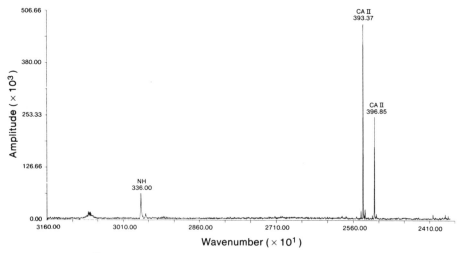

Fig. 26.   Plasma emission spectrum of 1.0-ppm $Ca(NO_3)_2$.

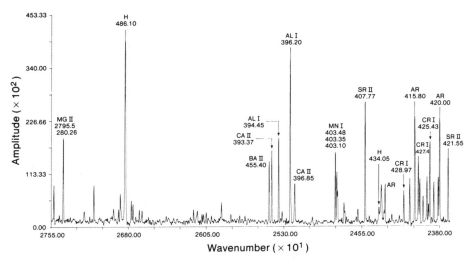

Fig. 27. Multielement argon plasma emission spectrum.

dergo emission in the plasma. The plasma thus naturally lends itself to multielement analysis very readily and, when coupled to an interferometer detection system, simultaneous multielement analysis is achievable. In Fig. 27 a portion of the emission spectrum, obtained with the plasma–interferometer system of a multielement mixture, is shown. Some major lines are identified on the figure, and a more complete list of the observed and literature elemental emission lines for the elements present as well as their concentrations is given in Table 7. This spectrum is an excellent example of the wide spectral coverage simultaneously available. For example, the manganese emission line at 259.5 nm and the barium emission line at 553.5 nm may be simultaneously monitored as well as all of the other spectral information present within the response range of the detector. In some cases, where the background or matrix emission is of much greater intensity than the analyte emission, the limited dynamic range problem previously discussed may exist. Now let us further consider this problem as it relates to the plasma emission source in the uv–visible spectral range.

Consider the case of trace aluminum determination in the presence of a high-concentration calcium matrix. Two sensitive aluminum neutral atom emission lines are very close to the calcium ion emission lines. However, at least a factor of 250 exists in their plasma emission sensitivity in favor of Ca. Optical or electronic filtering is out of the question due to their close proximity. The only alternatives rest in removing the calcium in a sample preanalysis step or in choosing a different, less-sensitive, alumi-

TABLE 7

Multielement Inductively Coupled Plasma
Emission with a 1P28 Photomultiplier Tube

Element	Conc. (ppm)	Observed wavelength (nm)	Literature wavelength[a] (nm)
MgI	10.0	285.18	285.21
MgII	10.0	280.25	280.26
		279.54	279.55
MnI	30.0	403.07	403.08
		403.32	403.31
		403.47	403.45
MnII	30.0	294.90	294.92
		293.92	293.93
		293.30	293.31
		260.55	260.57
		257.59	257.61
FeI	50.0	374.93	374.95
		374.58	374.56
		373.71	373.71
		373.50	373.49
		371.99	371.99
		361.74	361.78
		358.11	358.12
		356.98	357.01
		356.71	356.74
		355.49	355.49
AlI	50.0	309.24	309.28
		308.20	308.22
BaI	0.4	553.46	553.55
BaII	0.4	493.35	493.41
		455.34	455.40
SrII	0.2	421.56	421.55
		407.76	407.78
CaII	0.1	396.86	396.85
		393.38	393.37

[a] From Zaidel' *et al.* (1970).

num line. In another case, in attempting to obtain quantitative information on calcium emission in the form of an analytical curve extending over several orders of magnitude, background lines were a problem. Without optical filtering, the 486.1-nm hydrogen emission line quickly became a dynamic range limitation in the concentration range of 0.10-ppm calcium. After optical isolation of the calcium ion emission lines, the argon line emission at 394.8 nm, midway between the ion emission lines, became a dynamic range limitation at calcium concentrations on the order of 0.01-ppm calcium. Typical detection limits using a plasma monochromator-based system are below 0.1 ppb. Thus the problem of dynamic range limitations may ultimately limit the sensitivity of the interferometer-based system for elements in a spectral position similar to that of calcium.

As with the flame source, some indication of the spectral distribution of noise in the interferometer–ICP-based system was sought. To gain this information one would ideally like to monitor the standard deviation of a small emission signal, both in the presence of and in the absence of a larger amplitude emission signal. Normally this is accomplished using two separate solutions, one with and one without the high concentration species. A novel approach used only one solution containing both analytes and employs the use of an optical filter that selectively eliminates emission from the high concentration species. Such an experiment was performed using the magnesium line emissions in the ultraviolet as the high-intensity signal and the calcium ion line emissions in the visible as the low-level signal. The magnesium was present at 100 ppm and the calcium at 0.10 ppm. Corning filter 0-54 was used to isolate the emission lines due to its 100% transmittance above 300 nm. Example spectra of the calcium ion emission lines at 393.4 and 396.8 nm with and without the magnesium emission detected are shown in Fig. 28 and 29. A factor of about ten exists between the Mg and Ca peak amplitudes, with Mg the more intense. The variances of the 393.4-nm Ca ion emission line are presented in Table 8. Under a variance ratio F test at $\alpha = 0.05$, the variances with and without the presence of Mg were found to be statistically equivalent. A similar test on the H 486.1-nm and Ar 420.1-nm emission lines also indicated no change in the peak variance due to the presence of Mg emission. This is definite evidence that at least at the detection-processing steps a strong spectral signal does not distribute significant amounts of noise across the base line and hence to other analyte lines. Note that this is similar to the situation found for flame emission sources.

A rather large number of elements have their most sensitive ICP emission lines in the 200–300 nm region (Winge *et al.*, 1979; Boumans and Bosvedd, 1979). Chromium, manganese, cadmium, silicon, magnesium, boron, tin, lead, zinc, and iron are among some of the elements that fall

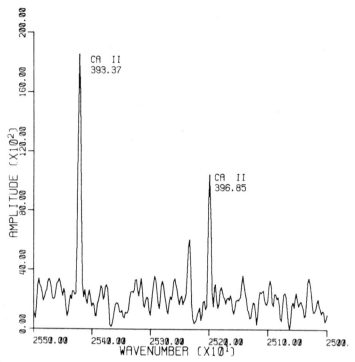

Fig. 28. Ca (0.1-ppm) ion emission lines in the presence of unfiltered 100-ppm Mg.

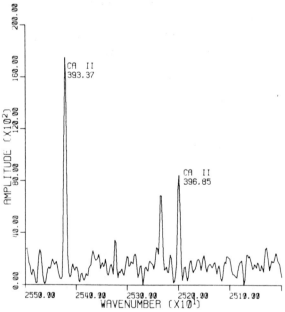

Fig. 29. Ca (0.1-ppm) ion emission lines in the presence of filtered 100-ppm Mg.

TABLE 8

Variance Data for Ca–Mg Study[a]

Ca 393.3-nm emission peak	Ca + Mg (no filter)	Ca + Mg (filter)
Mean	1.97765×10^4	1.86731×10^4
Variance	1.0240×10^7	4.3482×10^6
Relative standard deviation (%)	16.19	11.16

[a] Ca = 0.1 ppm, Mg = 100 ppm.

into this class. Although the current interferometer system does not respond well below about 240 nm, several important lines can be measured in the region from 240 to 300 nm using the solar blind photomultiplier tube. With this detector the plasma spectral background is completely clean of any argon- or solvent-generated emissions. The partial emission spectrum in the solar blind regions (316.6–253.1 and 253.1–210.9) of Table 2 of a solution containing tin, lead, chromium, boron, iron, silicon, magnesium, and manganese is shown in Fig. 30. The concentrations utilized and lines identified are shown in Table 9. These are convenient concentration levels and are not indicative of detection limits, which would be considerably lower.

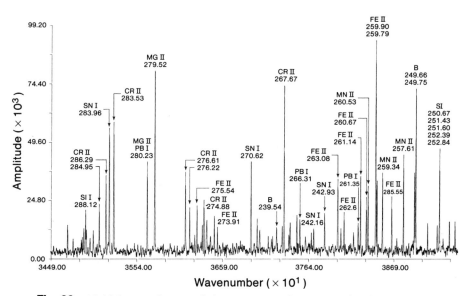

Fig. 30.   Multielement plasma emission spectrum using a solar blind PMT detector.

TABLE 9

Multielement Inductively Coupled Plasma Emission with a Solar Blind
Photomultiplier Tube

Element	Conc. (ppm)	Observed wavelength (nm)	Literature wavelength[a] (nm)	Element	Conc. (ppm)	Observed wavelength (nm)	Literature wavelength[a] (nm)
CrII	20	286.29	286.26	FeII	150	275.54	275.57
		294.95	284.98			274.62	274.65
		283.53	383.56			273.91	273.95
		276.61	276.66			263.08	263.11
		276.22	276.26			262.80	262.83
		274.88	274.90			262.53	262.55
		267.67	267.72			261.72	261.76
BI	100	249.75	249.77			261.14	261.18
		249.66	249.68			260.67	260.71
		239.54	239.51			259.90	259.94
PbI	200	280.23	280.20			259.79	259.84
		266.31	266.32			258.55	258.59
		261.35	261.42	MnII	10	260.52	260.57
SnI	200	283.96	284.00			259.34	259.37
		270.62	270.65			257.57	257.61
		254.55	254.66	MgII	4	280.23	280.26
		248.32	248.34			279.52	279.55
		242.93	242.95				
		242.16	242.17				
SiI	250	288.12	288.16				
		252.84	252.85				
		252.39	252.41				
		251.60	251.92				
		251.43	251.43				
		250.67	250.69				

[a] From Zaidel' et al. (1970).

Note the resolution capability available in this spectral region, especially with respect to the boron doublet at 249.66 and 249.75 nm and the iron doublet at 259.84 and 259.94 nm. Even at this resolution, 0.0483 nm, problems do occur with respect to the spectral overlap of aliased frequencies. For example, the silicon emission line at 251.61 nm is aliased directly on top of the 254.66-nm emission line of tin. The silicon line aliases to 254.65 nm. With the current resolution, these peaks are not resolvable, but an increase by a factor of five in resolution, to 1.55 cm^{-1}, or 0.01 nm, would adequately resolve these peaks. Aliasing may or may not be a serious problem depending on the relative intensities of the two lines,

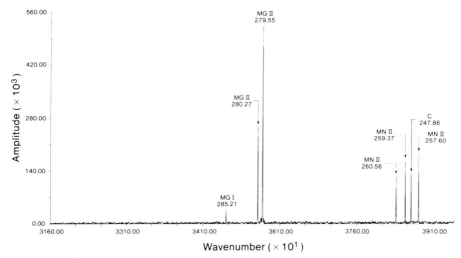

Fig. 31.   Plasma emission spectrum of Mn (4.4 ppm) and Mg (20 ppm).

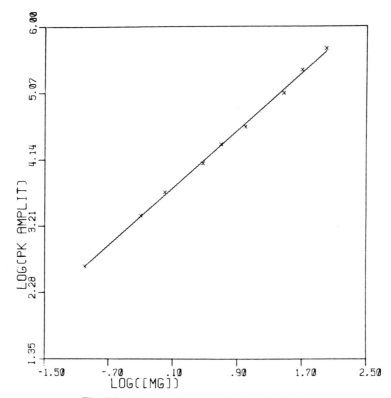

Fig. 32.   Mg analytical curve for data in Fig. 31.

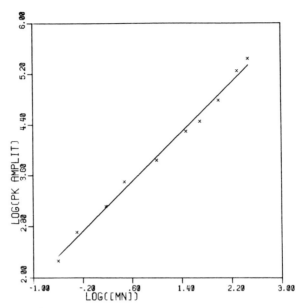

Fig. 33. Mn analytical curve for data in Fig. 31.

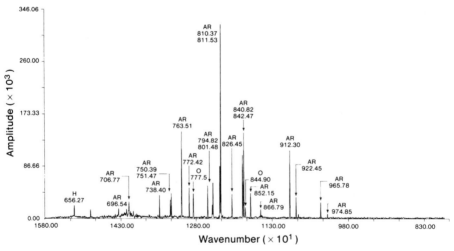

Fig. 34. Plasma background emission spectrum using a silicon photodiode detector.

their wavelength separation, and whether or not any other analytically useful lines are available for the quantitative determination. This spectrum and Table 9 clearly illustrate the potential power of the interferometer–ICP system for precise qualitative identification of sample composition.

Manganese and magnesium were used to demonstrate simultaneous quantitative determinations. A spectrum resulting from an aqueous solution containing 20-ppm Mg and 4.4-ppm Mn is shown in Fig. 31. The carbon line at 247.86 was due to a carbon-based contaminant in the argon gas supply. Analytical curves were simultaneously determined for Mg ranging

TABLE 10

Argon Plasma Background Emission
as Observed with a Silicon
Photodiode Detector

Element	Observed wavelength (nm)	Literature wavelength[a] (nm)
ArI	696.59	696.54
	706.75	706.72
	738.48	738.40
	750.45	750.39
	751.54	751.47
	763.61	763.51
	772.49	772.42
	794.88	794.82
	801.64	801.48
	810.55	810.37
	811.69	811.53
	826.58	826.45
	840.93	840.82
	842.57	842.47
	852.27	852.15
	866.95	866.79
	912.43	912.30
	922.67	922.45
	966.01	965.78
	978.59	978.45
H	656.27	656.29
O	777.50	777.41
	844.90	844.64

[a] From Zaidel' et al. (1970).

from 100 to 0.05 ppm and Mn ranging from 300 to 0.25 ppm (see Fig. 32 and 33). Although this is a very simple example, the potential for simultaneous quantitative multielement analysis is clear.

Finally, the interferometer system was used to measure the complex plasma spectral background emission in the near-ir. This region is readily accessible with the interferometer when it is coupled to the Si photodiode detector. The spectrum of the plasma in this region is shown in Fig. 34 and the lines identified in Table 10.

The qualitative and quantitative spectrochemical data presented in this section clearly illustrate the potential capability of a Fourier-transform-spectrometer-based measurement system for atomic emission spectrochemical measurements. The wide-ranging and detailed simultaneous wavelength coverage is matched only by the photographic plate. While suffering some limitations in intensity, dynamic range, and detection limits, measurements can be made at the sub-ppm ranges for most elements using this interferometer–ICP combination. Work is currently underway in our laboratory to optimize beam-splitter performance for ultraviolet performance and to improve resolution by at least a factor of four. We feel that these improvements will result in an almost ideal measurement system for simultaneous multielement analysis in the most general sense.

REFERENCES

Aldous, K. M., Browner, R. F., Dagnall, R. M. and West, T. S. (1970). *Anal. Chem.* **42,** 939.
Boumans, P. W. J. M., and de Boer, F. J. (1975). *Spectrochim. Acta* **30B,** 309.
Boumans, P. W. J. M., and Bosveld, M. (1979). *Spectrochim. Acta* **34B,** 59.
Chester, T. L., Fitzgerald, J. J., and Winefordner, J. D. (1976). *Anal. Chem.* **48,** 793.
Dahlquist, R. L., and Knoll, J. W. (1978). *Appl. Spectrosc.* **32,** 1.
Edmonds, T. E., and Horlick, G. (1977). *Appl. Spectrosc.* **31,** 536.
Filler, A. S. (1973). *J. Opt. Soc. Am.* **63,** 589.
Gerstenkorn, S., Luc, P., Perrin, A., and Chaurille, J. (1977a). *Astron. Astrophys.* **58,** 255.
Gerstenkorn, S., Luc, P., and Perrin, A. (1977b). *J. Mol. Spectrosc.* **64,** 56.
Greenfield, S., McGeachin, H. McD., and Smith, P. B. (1976). *Talanta* **23,** 1.
Harris, J. J., Lytle, F. E., and McCain, T. C. (1976). *Anal. Chem.* **48,** 2095.
Harwit, M., and Tai, M. H. (1977). *Appl. Opt.* **16,** 3071.
Hirschfeld, T. (1976). *Appl. Spectrosc.* **30,** 68.
Horlick, G., and Malmstadt, H. V. (1970). *Anal. Chem.* **42,** 1361.
Horlick, G., and Yuen, W. K. (1975). *Anal. Chem.* **47,** 775A.
Horlick, G. and Yuen, W. K. (1978). *Appl. Spectrosc.* **32,** 38.
Kahn, F. D. (1959). *Astrophys. J.* **129,** 518.
Knacke, R. F. (1978). *App. Opt.* **17,** 684.
Luc, P., and Gerstenkorn, S. (1978). *Appl. Opt.* **17,** 1327.
Lytle, F. E. (1974). *Anal. Chem.* **46,** 545A.
Malmstadt, H. V., Enke, C. G., Crouch, S. R., and Horlick, G. (1974). "Optimization of Electronic Measurements". Benjamin, New York.

Marshall, A. G., and Comisarow, M. B. (1975). *Anal. Chem.* **47,** 491A.

Pickett, E. E., and Koirtyohann, S. R. (1969). *Anal. Chem.* **41,** 28A.

Plankey, F. W., Glenn, T. H., Hart, L. P., and Winefordner, J. D. (1974). *Anal. Chem.* **46,** 1000.

Tai, M. H., and Harwit, M. (1976). *Appl. Opt.* **15,** 2664.

Talmi, Y. (ed.) (1979). "Multichannel Image Detectors", ACS Symp. Ser. 102. American Chemical Society, Washington D.C.

Treffers, R. R. (1977). *Appl. Opt.* **16,** 3103.

Winefordner, J. D., and Haraguchi, H. (1977). *Appl. Spectrosc.* **31,** 195.

Winefordner, J. D. *et al.* (1976). *Spectrochim. Acta* **31B,** 1.

Winge, R. K., Fassel, V. A., and Kniseley, R. N. (1977). *Spectrochim. Acta* **32B,** 327.

Winge, R. K., Peterson, U. J., and Fassel, V. A. (1979). *Appl. Spectrosc.* **33,** 206.

Yuen, W. K., and Horlick, G. (1977). *Anal. Chem.* **49,** 1446.

Zaidel', A. N., Prokof'ev, V. K., Raisku, S. M., Slavngi, V. A., and Shreider, E. Ya. (1970). "Tables of Spectral Lines". Plenum Press, New York.

3

DOUBLE MODULATION FOURIER TRANSFORM SPECTROSCOPY

Laurence A. Nafie

Department of Chemistry
Syracuse University
Syracuse, New York

D. Warren Vidrine

Nicolet Instrument Corporation
Madison, Wisconsin

I. INTRODUCTION

The measurement of difference spectra is highly important for the study of subtle phenomena at the molecular level. This is particularly true for vibrational spectra where differences occur in a rich manifold of transitions containing structural information from virtually all portions of a molecule. Moreover, difference spectra perform a selective function by focusing attention on the regions of greatest contrast between two spectra and drawing out delicate features such as small frequency shifts in spectral bands, which otherwise might not be apparent.

FOURIER TRANSFORM
INFRARED SPECTROSCOPY, VOL. 3

Currently, there are two conceptually distinct approaches to the measurement of difference spectra. The first is a dual-spectrum approach, where the two spectra leading to the difference spectrum are individually available in the course of the measurement process. This involves either consecutive or simultaneous recording of the two parent spectra to a sufficiently high level of precision, followed by the subtraction of one from the other. The second is a differential-spectrum approach, where the measurement process follows only a signal modulating between two spectroscopic states and where the spectra associated with these two states, representing the maximum and the minimum of the modulation cycle, are not directly measured. In this chapter we shall refer to the first method as difference spectroscopy and to the second as differential spectroscopy, although both methods result in a spectrum representing the difference between two spectral states.

The straightforward way to obtain a difference spectrum is to record two parent spectra separately, using a standard dispersive spectrometer, and then subtract one spectrum from the other. This approach suffers from a number of drawbacks, the most significant being that the parent spectra are measured during two different time periods and that different wavelength points in each spectrum are also measured at different times. Two separate advances in spectroscopic instrumentation and measurement techniques have made possible the elimination of either one or the other of these two drawbacks but not both in one measurement process. In particular, if one employs the approach of differential spectroscopy modulating rapidly between the parent spectroscopic states, the drawback of measuring the two parent states of different periods in time is eliminated. On the other hand, if Fourier transform spectroscopy is employed, all spectral points in the parent spectra are measured simultaneously. Both of these improvements have, in their own way, dramatically improved the accuracy, precision, and quality of difference spectra, making possible the observation of exceedingly small spectral differences.

The first of these improvements, which employs rapid modulation and differential measurement, embodies the so-called ac advantage. This states that a small difference signal can be measured more accurately as the amplitude of a periodically varying ac signal than as the difference between two time-independent dc signals. Essentially, the ac signal represents a large number of oscillations between two states. Measurement of the ac-signal amplitude involves averaging over these oscillations, which reduces random noise contributions and confers a distinct advantage over a single dc measurement. The measurement of such ac signals is best achieved with a tuned lock-in amplifier that has the capability of rejecting interfering noise, having frequency components that are either outside the bandpass filter or else at the modulation frequency but out of phase with the signal being measured. The availability of high-quality lock-in amplifiers, as well as light choppers and electrooptic devices for producing a modulated signal, has made differential modulation spectroscopy an attractive means for obtaining quality difference spectra.

The other method for the improvement of difference spectra is Fourier transform (FT) spectroscopy (Bell, 1972; Griffiths, 1975). The requirements for obtaining quality FT-difference spectra have been discussed by a number of authors (Strassburger and Smith, 1979; Hirschfeld, 1976, 1979; Griffiths, 1977, 1978a,b), and a number of specific advantages arise with this approach. These include Jacquinot's advantage of increased throughput (Jacquinot, 1960) and Fellgett's advantage of multiplex detection (Fellgett, 1958), both of which permit either shorter measuring times or, more importantly, higher spectral quality under the same measurement times when compared to a corresponding dispersive spectrometer. Another important advantage of using FT spectroscopy is Connes's advantage (Connes and Connes, 1966), which arises from the continuous calibration of infrared spectral frequencies by a laser fringe counting technique. This permits the precise alignment of the frequencies of consecutively measured spectra. Thus, even though difference spectra in FT spectroscopy are obtained by the subtraction of two spectra, a major source of error in such a process is reduced by Connes's advantage (Hirschfeld, 1975). As a consequence, it is possible to obtain high-precision vibrational difference spectra in the infrared using FT-IR spectrometers, and a number of specific examples have appeared in the literature (Koenig, 1975; Gendreau and Griffiths, 1976; Painter and Koenig, 1977; D'Esposito and Koenig, 1978).

The question that now arises is whether the modulation methods of differential spectroscopy can be incorporated into FT spectroscopy to yield a measurement process that simultaneously overcomes the disadvantages of measuring the two parent spectra at different times and measuring different wavelengths at different times. Such a method has been recently developed (Nafie and Diem, 1979a; Nafie *et al.*, 1979; Nafie and Vidrine, 1979) and is referred to here as double modulation Fourier transform (DM-FT) spectroscopy. The basic thrust of this new method is to combine the ac advantage and the FT advantages into a single measurement process. This is achieved by modulating the sample absorption between the desired extremes, as in ordinary differential spectroscopy, while carrying out the spectral measurements with an FT spectrometer. The basic requirement of DM-FT spectroscopy is that the differential modulation frequency f_m be at least an order of magnitude higher than the highest of the

Fourier interferogram frequencies. Thus there are two different types of modulation present, namely, the usual band of Fourier frequency modulations caused by the interferometer and the high-frequency differential modulation. As a result, there are *two* interferograms present in the detector signal. One is the usual interferogram representing the average transmission of the sample and the other is a new interferogram that is carried by the modulation frequency f_m and represents the desired difference spectrum. The transmission interferogram may be Fourier transformed in the usual way but the differential interferogram which is initially present as sideband modulations on the carrier frequency f_m, as shown in Fig. 1, must first be demodulated by a lock-in amplifier tuned to f_m. If the time constant of the lock-in is set to a sufficiently low level, a demodulated differential interferogram is available at the output of the lock-in. This may be Fourier transformed in essentially the same manner as the transmission interferogram to yield the difference spectrum.

The idea of superimposing an interferogram on a higher modulation frequency has been practiced before in FT spectroscopy. The early slow-scanning far-infrared FT spectrometers used a mechanical light chopper, operating at a few hundred hertz, to better define the slowly varying interferogram signal that appeared at the output of a lock-in amplifier tuned to the chopping frequency (Griffiths, 1975). In addition, a group of French workers (Russel *et al.*, 1972) pointed out the possibility of combining high-frequency polarization modulation with FT spectroscopy in order to measure circular dichroism spectra. From a historical point of view this chapter represents the first mention of double modulation FT differential spectroscopy from a general standpoint.

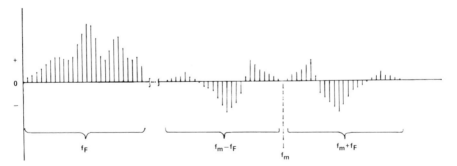

Fig. 1. A discrete display of the intensity of the frequency components of the detector signal. The ordinary transmission intensities appear at Fourier frequencies f_F to the left. The double-modulated difference intensities appear as side bands symetrically disposed about the high-frequency carrier frequency f_m.

Thus far we have made no explicit reference to double-beam methods, which are widely available in commercial absorption spectrometers and which yield the difference spectrum of the sample beam minus the reference beam. According to our classification scheme, double-beam dispersive absorption spectroscopy must be regarded as a form of differential spectroscopy since individual sample and reference spectra are not normally measured. In FT spectroscopy, double-beam absorption is available only as difference spectroscopy since the sample and reference are measured consecutively even though there may be separate optical paths for sample and reference. In order to obtain true differential double-beam FT absorption, a rapid modulation is required between sample and reference as each interferogram is scanned. However, this is simply a specific form of DM-FT spectroscopy as previously described.

At this point it is important to mention that recently another method has been developed for the double-beam measurement of FT spectra. Referred to as dual-beam Fourier transform spectroscopy (Kuehl and Griffiths, 1978), this method is the optical analog of DM-FT spectroscopy. In this approach, the beam, which normally returns to the source from the interferometer, is recovered and treated in a manner equivalent to the sample beam. Since the two beams have complementary Fourier modulation (90° phase difference), only Fourier signals representing the difference in intensity of the two beams are present when the beams are focused together at the detector. The resulting differential interferogram is the same as the demodulated differential interferogram obtained in DM-FT spectroscopy. Each of these two methods has its own advantages and disadvantages, and an investigator should be aware of both approaches when considering the design of a differential experiment.

Finally, we want to stress that DM-FT spectroscopy is quite general, requiring only transmission modulation at a frequency significantly higher than the Fourier frequencies. For rapid-scan FT spectrometers, this requires f_m to be in the tens of kilohertz range. This is readily achieved with polarization modulation for measuring samples that depend on the polarization of the beam. Other methods include sample absorption modulation achieved either by switching the beam between two different samples or by actually changing the sample's absorption through some external perturbation, such as an electric field. It may also be possible to study kinetic events by the periodic initiation of photochemical or electrochemical events in a cycle that includes relaxation back to equilibrium.

In this chapter we shall first describe the mathematical basis of DM-FT spectroscopy, which will be followed by a discussion of the instrumental aspects of this method. Different types of differential modulation will then be described and, as an example of this technique, we shall conclude by describing experiments measuring vibrational circular dichroism (Nafie *et al.*, 1976; Nafie and Diem, 1979b), which is the differential absorbance of left versus right circularly polarized light by optically active molecules. It is our aim in this chapter to encourage the use of DM-FT as a sensitive new method for obtaining difference spectra in infrared spectroscopy.

Before concluding the introduction we should like to mention two other methods that hold potential for the measurement of vibrational circular dichroism (VCD). The first is an application of dual-beam FT-IR spectroscopy (Griffiths, 1978c) in which one beam is left circularly polarized and the other beam is right circularly polarized. Any VCD in the sample then results in a small differential interferogram at the detector. Considerable care is required to balance the two beams to the required level of precision, but detector saturation is not a problem even at high light levels because the VCD interferogram is small and the large transmission interferogram is cancelled.

The other method for measurement of circular dichroism, (Dignam and Baker, 1980) which is based on the concept of a scanning polarization interferometer (SPI), was originally proposed by Martin and Puplett (1969). In this approach, a polarizing wire grid beam splitter is used together with rooftop corner reflecting mirrors at the moving and stationary mirror positions of the interferometer. Two unique features result from this design. First, all of the light goes on toward the sample in one beam and none is returned to source. Second, the light is not intensity modulated but is rather polarization modulated. As a result, each wavelength experiences typically thousands of phase retardation cycles. Consequently, the SPI may become an important method for the measurement of VCD, especially at very low vibrational frequencies (< 500 cm^{-1}) where the creation of left and right circularly polarized light with a photoelastic modulator is difficult and has not yet been achieved. It should also be mentioned that a *single*-beam SPI can be configured to act as a dual-beam interferometer.

II. THEORETICAL DESCRIPTION

Consider the two-parent transmission spectra $I_\alpha(\bar{\nu})$ and $I_\beta(\bar{\nu})$, where $\bar{\nu}$ represents the wavenumber frequency of the infrared radiation. If one modulates between these two spectra at the frequency f_m, the total intensity in watts reaching the detector as a function of the time and wavenumber is

$$I(\bar{\nu},t) = I_{dc}(\bar{\nu}) + I_{ac}(\bar{\nu}) \sin 2\pi f_m t \qquad (1)$$

where the average transmission I_{dc} and the difference transmission I_{ac} are given by

$$I_{dc}(\bar{\nu}) = \tfrac{1}{2}[I_\alpha(\bar{\nu}) + I_\beta(\bar{\nu})] \tag{2}$$

$$I_{ac}(\bar{\nu}) = \tfrac{1}{2}[I_\alpha(\bar{\nu}) - I_\beta(\bar{\nu})] \tag{3}$$

We have expressed the modulation cycle as a sine wave for the sake of simplicity and convenience, but other modulation waveforms may be used and are not intentionally excluded. Although the difference spectrum given by $I_{ac}(\bar{\nu})$ is an adequate description, one more compatible with theoretical interpretation in terms of molecular properties is expressed in terms of the decadic absorbance A defined in relation to the transmission spectra by

$$I_{dc}(\bar{\nu}) = \tfrac{1}{2}I_0(\bar{\nu})[10^{-A_\alpha(\bar{\nu})} + 10^{-A_\beta(\bar{\nu})}] \tag{4}$$

$$I_{ac}(\bar{\nu}) = \tfrac{1}{2}I_0(\bar{\nu})[10^{-A_\alpha(\bar{\nu})} - 10^{-A_\beta(\bar{\nu})}] \tag{5}$$

where $I_0(\bar{\nu})$ is the throughput spectrum of the instrument. If the ratio of I_{ac} to I_{dc} is formed, one obtains an expression related to the absorbance difference spectrum, or, more specifically,

$$I_{ac}(\bar{\nu})/I_{dc}(\bar{\nu}) = \tanh\{\tfrac{1}{2}(\ln 10)[\Delta A_{\beta\alpha}(\bar{\nu})]\} \tag{6}$$

where $\Delta A_{\beta\alpha}(\bar{\nu}) = A_\beta(\bar{\nu}) - A_\alpha(\bar{\nu})$. For small values of ΔA the hyperbolic tangent may be replaced by its argument with little error (e.g., there is only 0.03% error in replacing $\tanh(x)$ by x when $x = 0.1$). Using this approximation we have

$$I_{ac}(\bar{\nu})/I_{dc}(\bar{\nu}) = 1.1513 \,\Delta A_{\beta\alpha}(\bar{\nu}) \tag{7}$$

Another advantage of obtaining the spectrum $\Delta A(\bar{\nu})$ rather than $I_{ac}(\bar{\nu})$ is that $\Delta A(\bar{\nu})$ does not depend on the instrument transmission profile I_0.

We shall now develop the corresponding set of equations for a Fourier transform spectrometer equipped for differential measurement. Since important steps in the measurement process take place after the detector converts radiant intensity $I(\bar{\nu})$ in watts to an electronic signal $B(\bar{\nu})$ in volts, we shall focus attention on electronic signals after the detector. The response function of the detector $R(\bar{\nu})$ in volts per watt then provides the connection between voltage and intensity by the relation

$$B(\bar{\nu}) = R(\bar{\nu})I(\bar{\nu}) \tag{8}$$

In general, the interferogram V from an FT spectrometer is expressed in terms of the position of the scanning mirror δ, and $V(\delta)$ is related to the transmission signal $B(\bar{\nu})$ by (Griffiths, 1978b)

$$V(\delta) = \int_0^\infty B(\bar{\nu}) \cos[2\pi\bar{\nu}\delta - \theta(\bar{\nu})] \, d\bar{\nu} \tag{9}$$

where $\theta(\bar{\nu})$ is a function describing the phase of the individual wave-number frequencies. If modulation between the two spectra $I_\alpha(\bar{\nu})$ and $I_\beta(\bar{\nu})$ is carried out, then from Eqs. (1), (8), and (9) the expression for the total interferogram becomes

$$V(\delta) = \int_0^\infty [B_{dc}(\bar{\nu}) + B_{ac}(\bar{\nu}) \sin 2\pi f_m t] \cos[2\pi\bar{\nu}\delta - \theta(\bar{\nu})] \, d\bar{\nu} \qquad (10)$$

If the detector signal is first sent through a low-pass electronic filter, which removes the high-frequency components containing f_m, one obtains the usual dc interferogram representing the average transmission, namely,

$$V_{dc}(\delta) = \int_0^\infty B_{dc}(\bar{\nu}) \cos[2\pi\bar{\nu}\delta - \theta_{dc}(\bar{\nu})] \, d\bar{\nu} \qquad (11)$$

where $\theta_{dc}(\bar{\nu})$ is the total dc-phase function including the electronic filtering. In order to obtain the ac interferogram the detector signal is first sent through a band-pass filter centered at f_m and then to a lock-in amplifier tuned to f_m. If the time constant of the lock-in (τ) is set sufficiently low, the ac interferogram will be present at the output. The time constant circuitry of the lock-in attenuates the higher Fourier frequencies more than the lower frequencies. This attenuation may be complex and depend on the exact nature of the time constant circuitry, but it may be approximated by a decreasing exponential of the form $\exp[-(2V\bar{\nu})\tau]$, where V is the velocity of the mirror in the interferometer and $2V\bar{\nu}$ the Fourier frequency for each wavenumber frequency. Consequently, the expression for the demodulated ac interferogram is

$$V_{ac}(\delta) = \int_0^\infty B_{ac}(\bar{\nu}) \exp[-(2V\bar{\nu})\tau] \cos[2\pi\bar{\nu}\delta - \theta_{ac}(\bar{\nu})] \, d\bar{\nu} \qquad (12)$$

Again $\theta_{ac}(\bar{\nu})$ incorporates all phase shifts along the ac path, including the lock-in amplifier and electronic filters.

Equations (11) and (12) represent interferograms that may be Fourier transformed and ratioed to yield

$$FT[V_{ac}(\delta)]/FT[V_{dc}(\delta)] \cong \exp[-(2V\bar{\nu})\tau](1.1513) \, \Delta A_{\beta\alpha}(\bar{\nu}) \qquad (13)$$

This expression differs from Eq. (7) only by the exponential attenuation factor of the lock-in amplifier. This factor can be determined by measuring either a suitable calibration spectrum or by carrying out a frequency response analysis of the lock-in amplifier time constant circuit.

III. INSTRUMENTAL CONSIDERATIONS

A. Basic Requirements

In order to carry out differential measurements using DM-FT spectroscopy, there are a number of prerequisites that must first be satisfied. As a guide to this discussion we shall first consider the general instrumental block diagram in Fig. 2, which shows both the optical and electronic pathways of the experiment. Light from the source is directed to the interferometer and the output beam is brought to a focus at the sample position as usual. At this point modulation of the sample absorption is introduced and the transmitted light is focused at the detector. After preamplification of the detector signal, two electronic pathways are available. One leads to a low-pass filter and Fourier transformation to yield the transmission spectrum. The other passes through a narrow band-pass filter centered at f_m (to remove the large Fourier modulations of the transmission interferogram) and then to a lock-in amplifier tuned to f_m whereupon the lock-in output is Fourier transformed to yield a difference spectrum.

The requirements of modulation frequency and detector response are the most crucial factors in the DM-FT method. These factors in turn are determined by the range of Fourier frequencies employed in the measurement. For a rapid scanning spectrometer, the mirror velocity V is typically 0.15 cm sec^{-1}. The resultant Fourier frequencies, given by the expression $2V\bar{\nu}$, then lie in the 120–1200 Hz range for the spectral limits 400–4000 cm^{-1}. Under these conditions, the modulation frequency must

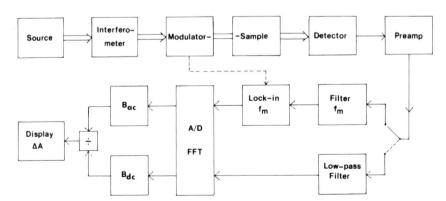

Fig. 2. Block diagram of the basic DM-FT instrumental setup. Double lines indicate optical signals and single lines indicate electronic signals. The modulator and sample may be one unit or separate units depending on the application.

be at least 10 kHz to allow adequate means to electronically filter and separate the transmission interferogram from the differential interferogram. This high modulation frequency then places a requirement of very fast response time on the infrared detector. Consequently, a solid-state infrared photodetector such as indium antinomide (InSb) or mercury–cadmium telluride (MCT) having response times of a microsecond or less must be employed.

Additional requirements involve the lock-in amplifier and the scan speed of the movable mirror in the interferometer. The lock-in must have a time constant setting in the 1-msec range in order to pass the upper range of Fourier frequencies near 1200 Hz. Longer time constants will begin to seriously diminish these higher frequencies, thus obscuring the upper range of vibrational frequencies. A further requirement, which is important for the sake of flexibility and frequency optimization, is that the interferometric mirror velocity should be variable. To obtain the highest-quality difference spectrum, the Fourier frequencies need to be optimized with respect to the high-frequency differential modulation, the lock-in time constant, and other electronic filters in the system. Furthermore, different spectral regions may require slightly different scan speeds to obtain the optimum results. However, to date detailed studies of the effect of scan speeds on DM-FT signal quality have not yet been carried out.

B. Phase Corrections

In order to Fourier transform an arbitrary interferogram to an infrared spectrum, one must determine what phase shifts occurred at the individual frequencies between their origin in the interferometer and their processing in the computer. Phase correction algorithms (Forman, 1966; Mertz, 1967) assume, in general, that only positive intensities are present in the experiment and that, as a result, the position of zero phase retardation for all Fourier frequencies can be determined by locating the point of maximum intensity in the interferogram.

For DM-FT measurements the assumption of positive intensities no longer holds since the interferogram, and hence the final spectrum, contains, in general, positive and negative intensities, as can be seen in Fig. 1. The point of zero phase retardation in the interferogram still represents the sum of all intensity contributions, but does not, in general, correspond to an intensity maximum. Consequently, the phase correction routine incorrectly chooses the point of zero retardation, and the phase corrections are erroneously determined.

The solution of this problem is to transfer the entire phase correction (zero point and wavelength dependence) from another spectrum, such as an appropriate calibration spectrum measured under conditions identical to the desired difference spectrum and where the calibration spectrum contains *only* positive or negative intensities. If such a monosignate calibration interferogram is not available, then the transmission interferogram of the sample may be used to obtain a close approximation to the true phase correction. The only errors introduced by using the transmission phase correction are due to phase shifts that occur in the f_m filter and the lock-in relative to the low-pass filter (Fig. 2). If it is assumed that phase shifts from these components are small relative to other sources of phase shift, then $\theta_{dc}(\bar{\nu}) \simeq \theta_{ac}(\bar{\nu})$ in Eqs. (11) and (12), respectively. We shall return to this point when discussing the results of our vibrational circular dichroism measurements.

C. Intensity Calibration

Once the interferogram $V_{ac}(\delta)$ has been successfully Fourier transformed to yield $B_{ac}(\bar{\nu})$ or $\Delta A(\bar{\nu})$, one still needs to calibrate the signs and intensities of the observed difference bands. The method used to secure such a calibration depends strongly on the nature of the differential modulation and differs for different types of experiments. Essentially, one needs to either measure the DM-FT spectrum of a sample having a known, and preferably large, difference spectrum obtained by an alternate method such as dispersive modulation spectroscopy or else a calibration spectrum that is also used in another difference spectroscopic procedure. In the latter case, one can measure the calibration spectrum using the DM-FT method, followed by the DM-FT difference spectrum of interest, where the relationship between these two measurements is known or can be readily determined.

IV. MODULATION TECHNIQUES

There are a number of conceptually distinct ways of creating high-frequency modulation of an FT transmission signal. For the present purpose we classify these as either polarization modulation or absorption modulation. Polarization modulation can be carried out effectively by using a photoelastic modulator (PEM) (Kemp, 1969; Cheng *et al.*, 1976) in which an isotropic optical element is periodically stressed by one or more piezoelectric crystals. This method is particularly important for nonlaser infrared applications since a PEM functions well even for wide acceptance angles. Absorption modulation is a more diverse classification. There are a wide variety of different approaches that can lead to the direct modulation of the absorption of the sample.

A. Polarization Modulation

The modulation of the polarization of a monochromatic light beam results naturally in the measurement of either circular dichroism (CD) or linear dichroism (LD). In the former, one is probing the difference in the sample absorption of left versus right circular polarized light while in the latter, one is comparing orthogonal states of linear polarized light in an oriented sample. The rapid sine-wave modulation of the polarization of a light beam is central to a very sensitive method of measuring CD and LD (Grosjean and Legrand, 1960; Velluz *et al.*, 1965). This method, known as the Grosjean–Legrand polarization modulation technique, revolutionized the measurement of CD, and to a lesser extent LD, and paved the way for the more recent discovery of infrared vibrational CD (Holzwarth *et al.*, 1974; Nafie *et al.*, 1975). The polarization modulation procedure to be described next corresponds to the natural extension of the Grosjean–Legrand polarization modulation method to FT-IR spectroscopy.

1. Circular Dichroism

If the polarization of a light beam is modulated sinusoidally between left and right circular polarization states, corresponding to oscillation of the retardation angle α of the PEM between $+90°$ and $-90°$ (see Section V.B for further details), a sinusoidal modulation in the intensity of the beam will arise after passing through an optically active sample having absorbances A_L and A_R for left and right circularly polarized light, respectively. In the Grosjean–Legrand method, the variation of the differential intensity depends on the sine of the retardation angle $\alpha(\bar{\nu},t)$ and the intensity of the CD, defined as $A_L(\bar{\nu}) - A_R(\bar{\nu})$. Accordingly, we can adapt the general intensity expressions in Eqs. (1)–(3) to these requirements to give

$$I(\bar{\nu},t) = I_{dc}(\bar{\nu}) + I_{ac}(\bar{\nu}) \sin[\alpha(\bar{\nu},t)] \tag{14}$$

$$I_{dc}(\bar{\nu}) = \tfrac{1}{2}I_0(\bar{\nu})[10^{-A_R(\bar{\nu})} + 10^{-A_L(\bar{\nu})}] \tag{15}$$

$$I_{ac}(\bar{\nu}) = \tfrac{1}{2}I_0(\bar{\nu})[10^{-A_R(\bar{\nu})} - 10^{-A_L(\bar{\nu})}] \tag{16}$$

Although the differential intensity depends on the sine of the retardation angle, this angle itself varies sinusoidally at the oscillation frequency f_m of the photoelastic modulator according to the equation

$$\alpha(\bar{\nu},t) = \alpha_0(\bar{\nu}) \sin 2\pi f_m t \tag{17}$$

where $\alpha_0(\bar{\nu})$ corresponds to the maximum retardation level achieved during the modulation cycle. The wavenumber dependence of α_0 is included since different frequencies $\bar{\nu}$ are retarded inversely proportional to their wavelength by the action of the modulator. Since α_0 also depends linearly on the maximum stress applied to the photoelastic optical element, higher stress levels are required to achieve quarter-wave retardation ($\alpha_0 = \pm 90°$) for increasingly longer wavelengths. In order to simplify the double sine dependence of the differential intensity [Eqs. (14) and (17)], we expand this sine factor in terms of odd-order spherical Bessel functions (which act as modulation efficiency factors for the modulation of the polarization at the frequency nf_m) as

$$\sin[\alpha_0(\bar{v})\,\sin\,2\pi f_m t] = 2 \sum_{n,\text{odd}} J_n[\alpha_0(\bar{v})]\,\sin\,2\pi n f_m t \qquad (18)$$

Since the ac signal must pass through a lock-in amplifier tuned to the frequency f_m, only the first term of the sum in Eq. (18) is measured and the higher odd harmonics are rejected. Therefore, Eq. (14) simplifies to

$$I(\bar{v},t) = I_{dc}(\bar{v}) + I_{ac}(\bar{v})2J_1[\alpha_0(\bar{v})]\,\sin\,2\pi f_m t \qquad (19)$$

The Bessel function $J_1[\alpha_0(\bar{v})]$ is then a measure of the efficiency of the modulator toward CD measurement at the wavenumber frequency $\bar{v}$.

If we now explicitly extend these expressions to FT measurement, the expressions for the dc and ac interferograms become

$$V_{dc}(\delta) = \int_0^\infty B_0(\bar{v})/2[10^{-A_R(\bar{v})} + 10^{-A_L(\bar{v})}]\cos[2\pi\bar{v}\delta - \theta_{dc}(\bar{v})]\,d\bar{v} \qquad (20)$$

$$V_{ac}(\delta) = \int_0^\infty B_0(\bar{v})/2[10^{-A_R(\bar{v})} - 10^{-A_L(\bar{v})}]2J_1[\alpha_0(\bar{v})]$$

$$\times\,\exp[(2V\bar{v})\tau]\cos[2\pi\bar{v}\delta - \theta_{ac}(\bar{v})]\,d\bar{v} \qquad (21)$$

When these two expressions are Fourier transformed and then ratioed, we obtain

$$\text{FT}[V_{ac}(\delta)]/\text{FT}[V_{dc}(\delta)] \cong 2J_1[\alpha_0(\bar{v})]\,\exp[-(2V\bar{v})\tau](1.1513)\,\Delta A_{LR}(\bar{v}) \qquad (22)$$

Consequently, in order to determine the CD spectrum, we need to determine the functions $J_1[\alpha_0(\bar{v})]$ and $\exp[-(2V\bar{v})\tau]$. Fortunately, the product of these two functions can be determined by a calibration procedure that is also used in the standard application of the Grosjean–Legrand CD method using a dispersive instrument (Nafie *et al.*, 1976). We shall describe the operation of this method in the context of FT spectroscopy in Section V.C.

2. Linear Dichroism

If the polarization of a light beam is modulated between orthogonal states of linear polarization, corresponding to the retardation angle oscillating between $+180°$ and $-180°$, it is possible to observe linear dichroism from an oriented sample where the polarization is alternately parallel and perpendicular to some designated direction in the sample. Linear dichroism spectra can be displayed as either $I_\perp(\bar{\nu}) - I_\parallel(\bar{\nu})$, $A_\perp(\bar{\nu}) - A_\parallel(\bar{\nu})$, or as the dichroic ratio $A_\perp(\bar{\nu})/A_\parallel(\bar{\nu})$. Some examples of LD studies in the literature include application to molecular crystals (Adamowicz and Fishman, 1972), stretched polymer fibers (Piseri et al., 1975), flowing polymer samples (Wada, 1972), stress induced in crystals (Boccara et al., 1973), and biological samples (Hofrichter and Eaton, 1976; Kusan and Holzwarth, 1976). In the work of Boccara et al. as well as Kusan and Holwarth, significant improvement in spectral quality and sensitivity was achieved through the use of the Grosjean–Legrand polarization modulation method where more than an order of magnitude improvement over nonmodulation methods was achieved.

Even further improvement should be available for infrared work by applying the DM-FT methods previously described. The basic equations are direct extensions of the previously described CD expressions, where now we have

$$I(\bar{\nu},t) = I_{dc}(\bar{\nu}) + I_{ac}(\bar{\nu}) \cos[\alpha(\bar{\nu},t)] \tag{23}$$

$$I_{dc}(\bar{\nu}) = \tfrac{1}{2}I_0(\bar{\nu})[10^{-A_\parallel(\bar{\nu})} + 10^{-A_\perp(\bar{\nu})}] \tag{24}$$

$$I_{ac}(\bar{\nu}) = \tfrac{1}{2}I_0(\bar{\nu})[10^{-A_\parallel(\bar{\nu})} - 10^{-A_\perp(\bar{\nu})}] \tag{25}$$

Note that the LD signal depends on the cosine of the retardation angle instead of the sine. This results in an expansion in even-order spherical Bessel functions as

$$\cos[\alpha_0(\bar{\nu}) \sin 2\pi f_m t] = J_0[\alpha_0(\bar{\nu})] + \sum_{n,even} 2J_n[\alpha_0(\bar{\nu})] \cos 2\pi n f_m t \tag{26}$$

The term $J_0[\alpha_0(\bar{\nu})]$ is just a dc contribution and the lowest-frequency ac contribution occurs for $n = 2$ at the frequency $2f_m$. This corresponds to the fact that the polarization alternates between parallel and perpendicular states twice each modulation cycle, since $+180°$ and $-180°$ are actually the same retardation state. Consequently, the lock-in amplifier must be tuned to $2f_m$ for LD measurement. The LD analog of Eq. (19) is then

$$I(\bar{\nu},t) = I_{dc}(\bar{\nu}) + I_{ac}(\bar{\nu})2J_2[\alpha_0(\bar{\nu})] \cos 4\pi f_m t \tag{27}$$

Following the previous analysis, this leads to

$$FT[V_{ac}(\delta)]/FT[V_{dc}(\delta)] \cong 2J_2[\alpha_0(\bar{\nu})] \exp[-(2V\bar{\nu})\tau](1.1513) \, \Delta A(\bar{\nu}) \tag{28}$$

where $\Delta A(\bar{\nu})$ is simply $A_\perp(\bar{\nu}) - A_\parallel(\bar{\nu})$.

B.   Absorption Modulation

Instead of modulating the polarization of the light beam to elicit a dichroic response from a static sample, it is possible to directly modulate the absorption of the sample itself. This can be accomplished by at least two basic methods. One involves modulating the sample absorption by an external periodic perturbation. The other method can be carried out using two samples in a double-beam arrangement and rapidly switching the transmission that reaches the detector back and forth between the two samples.

1.   External Periodic Perturbation

Double modulation FT experiments in the area of external periodic perturbation include the application of any periodic perturbation that alters the absorption strength of the sample. Examples of possible perturbations are high-frequency electric or magnetic fields, temperature modulation, stress modulation by piezoelectric crystals, electrochemical modulation, and laser pulse modulation. If a rapid scan interferometer is used, the requirement for generating enough sample modulation at a sufficiently high frequency becomes a serious problem, except possibly for the case of stress modulation. Consequently, it may prove necessary to use slower scan velocities or restrict measurements to lower frequency vibrations or both in order to obtain separable double modulation.

In several of these methods, there is a distinct possibility for probing the relaxation rate of kinetic events triggered by the external perturbation. Two examples are the modulation of the cell potential in an electrochemical sample or the repetitive firing of high-power laser pulses. In both cases relaxation rates, probed via the sample's FT-IR difference spectrum, could be ascertained by varying the period of the perturbing oscillation cycle. As the perturbation frequency is changed, the sample will be able to complete varying amounts of its excitation–relaxation cycle, and from the magnitude and sign of the differential spectrum, the temporal profile of this cycle could be determined. At very high frequencies the sample will not have a chance to respond between successive cycles, and at low frequencies the excitation–relaxation cycle will occupy only a small fraction of the modulation period. Therefore, when the modulation period is close in duration to the relaxation time, the maximum differential effects will be observed.

It is important to note that the observation of DM-FT effects with external perturbations does not require the sample to return completely to the unperturbed state. This is due, in part, to the high sensitivity of the DM-FT method, which is capable of observing differential signals that are only one part in 10^4 or 10^5 of the total transmission level.

2. Double-Beam Switching

Double-beam switching is conceptually straightforward and represents an example of "real-time" double-beam modulation spectroscopy using an FT-IR spectrometer. It simply amounts to switching the transmitted beam between two samples much faster than any of the observed Fourier frequencies. As in all previous examples of DM-FT spectroscopy, a lower limit on the speed of this switching modulation is determined by the band of Fourier frequencies and hence the mirror scanning velocity.

We shall describe, rather generally, two different methods of achieving the switching modulation. One method is an optomechanical approach where the ir beam is alternately directed to one sample and then the other in a square-wave modulation cycle. This could be achieved by a multislotted rotating sector mirror operating at high revolution speeds, although achieving a switching rate higher than 1 kHz would be difficult. A variation of this method is to pass light through both samples simultaneously and then modulate the angle of a focusing mirror so that first one sample focus and then the other falls on the detector surface.

A second method is a switching process, which is achieved using a PEM; hence the modulation frequency presents no problem. In this method, equal amounts of light are simultaneously passed through both samples. The two beams are then linearly polarized orthogonal to one another and passed through a PEM and a second polarizer. If the axes of the three polarizers and the PEM are properly aligned and if the PEM is undergoing half-wave retardation ($\pm 180°$), then the transmission of the two beams will each vary sinusoidally between zero and full transmission but they will differ in phase by 90° (as sine and cosine), and sinusoidal switching will be achieved. A drawback of this method relative to the mechanical switching methods is the presence of the $J_2[\alpha_0(\bar{\nu})]$ dependence of the difference spectrum, due to the fact that all wavelengths are not modulated between $+180°$ and $-180°$, the exact value of $\pm\alpha_0$ depending on $\bar{\nu}$. The equations used to describe this method are the same as those used to describe linear dichroism. The two beams in this experiment may represent either separate optical paths and separate samples or one beam and a partitioned sample cell with orthogonal polarizers affixed to the two cell halves.

V. VIBRATIONAL CIRCULAR DICHROISM

A. Introduction

Vibrational circular dichroism (VCD) together with its Raman scattering analog, known as Raman optical activity (ROA), comprise a new field of spectroscopy called vibrational optical activity (VOA). In both VCD and ROA one measures the difference in the response of a sample comprised of optically active molecules to left versus right circular polarized incident radiation. The principal advantage of VOA measurement is its three-dimensional sterochemical sensitivity, which is superimposed on the structurally rich set of vibrational transitions available in infrared and Raman spectroscopy. Several recent reviews are available (Nafie and Diem, 1979b; Stephens and Clark, 1979; Barron, 1978, 1979) that describe various aspects of VOA and survey the work carried out in this field to date.

The major challenge in the measurement of VCD arises from the weak magnitude of the effect. Typical ΔA_{LR} values are four to five orders of magnitude smaller than the absorbance (A_L or A_R) for the sample. This currently limits the magnitude and spectral range over which VCD can be measured to signals larger than 10^{-6} absorbance units and to the higher-frequency vibrational transitions, where instrument performance is generally better (e.g., sources, detectors, and PEM efficiency).

In this section we shall describe the details of experiments in which double modulation FT spectroscopy is applied to the measurement of VCD in the carbon–hydrogen stretching region (Nafie *et al.*, 1979). These efforts are the first steps toward the overall goal of using DM-FT to improve the sensitivity limit and to extend the spectral range of current VCD methods.

B. Instrumental Details

1. Optical Layout

We begin our description of the FT-VCD setup with the optical details. In Fig. 3 a top view of the Nicolet-7199A interferometer layout is shown. Light from one of two sources S1 or S2 is collected by the two-position mirror MF1 and directed *via* aperature A1 and the two fixed mirrors M1 and M2 into the interferometer. In our experiments we used a water-cooled glower source operating near 1200°C located at S2. The beam next encounters a germanium-coated KBr beam splitter BSIR, which directs the beam to the scanning mirror at M3 and the fixed mirror at M4. After

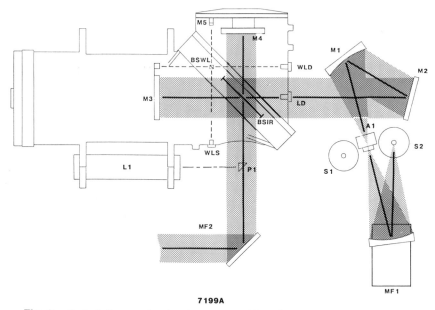

Fig. 3. Optical diagram showing the source and interferometer components. Also depicted are the laser reference and the white light optical components. See the text for detailed description.

recombination and interference at BSIR, the beam is reflected toward the sample by the mirror MF2. Also shown in Fig. 3 is the He–Ne laser fringe counting optics (Connes's advantage), which originates with the laser at L1, then concentrically shares the interferometer optics with the ir beam, and is detected at LD. Collinear He–Ne laser light remains with the ir beam into the sample and detector area to aid optical alignment. In addition, the white light reference source, beam splitter, and detector are shown. These elements initiate the recording of interferogram data by providing a sharp burst of white light at WLD at a predetermined position of mirror M3.

The optical layout in the region of the sample and the PEM is featured in Fig. 4. The beam from the interferometer (labeled "modulator") is focused by two mirrors to a BaF_2 wire grid, linear polarizer, a ZnSe PEM, and a sample cell with CaF_2 windows. The polarizer is set at 45° with respect to the orthogonal stress axes of the PEM so that equal intensities of polarized light lie along these axes. As shown in Fig. 5, the ZnSe optical element of the PEM is alternately stretched and compressed along these axes by attached quartz piezoelectric transducers (PZT). As a result, the indices of refraction of the ZnSe element along the stress axes vary sinusoidally with respect to one another causing the components of polarized

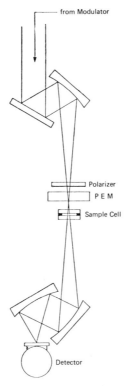

from Modulator

Polarizer

P E M

Sample Cell

Detector

Fig. 4. Optical diagram showing the top view of the region between the interferometer and the detector for VCD measurement.

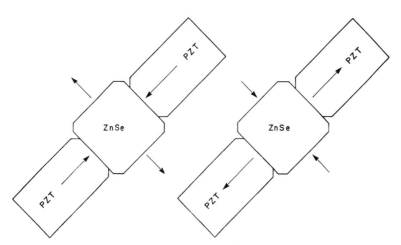

Fig. 5. Stress axes depicted for a ZnSe optical element of a PEM for the compression and expansion halves of the modulation cycle. The principal stress axis lies along the line defined by the PZT crystals and some stress alleviation takes place as shown in the orthogonal direction.

light along these axes to be phase advanced or retarded with respect to each other. When the phase retardation limits are $\pm 90°$, modulation between right and left circular polarization is achieved. The VCD sample is placed directly after the PEM to ensure that the polarization of the ir beam is not altered before absorption by the sample occurs. After the sample, the beam is focused on a liquid-nitrogen-cooled InSb solid-state photodetector.

2. Electronic Components

The electronic layout was described in general terms in Section III.A, and we shall provide details here that are more specific to our VCD application. The detector signal is amplified in the vicinity of the detector by a matched preamplifier. The electronic filtering is carried out by a series of computer-selectable high- and low-pass filters, which have a double pole roll-off. In general, the band of Fourier frequencies is flanked on the high and low side by these filters, which can be optimized for each of the 63 computer-selectable mirror velocities between 0.05 and 4.0 cm sec^{-1}. The lock-in amplifier employed in the measurements is by Princeton Applied Research Corp., model 124A. This lock-in is highly versatile, having a wide variety of gains, time constant settings, and tuned filter options. The reference channel of the lock-in obtains its signal directly from the control circuitry of the PEM. In the experiments to be described, the ZnSe PEM used was constructed at Syracuse University (Diem *et al.*, 1977, 1978). After the lock-in, the signal passes into the A/D converter and on to a Nicolet 1180 minicomputer system, having a 40K × 20-bit word memory. The digitized interferogram is then phase corrected and Fourier transformed using the fast Fourier transform algorithm (Forman, 1966).

Figure 6 is a photograph of the instrument setup. The polarizer, PEM, and sample can be seen mounted at one of the output ports of the interferometer. Several mirrors then bring the beam to a focus at the InSb detector, which is in a cylindrical Dewar toward the rear of the optical layout. The control electronics for the PEM and the band-pass filter centered at f_m together with the PARC 124A lock-in amplifier are located on the small table in front of the interferometer. A cable running between the PEM and the small table provides control for the PEM and a reference for the lock-in.

C. Calibration Spectra

Earlier we discussed the need to perform suitable calibration measurements in order to properly determine the phase correction for the differential interferogram and to establish the intensity scale of the observed dif-

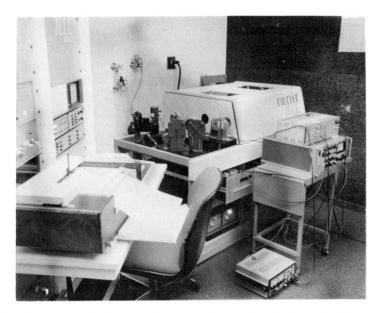

Fig. 6.   Photograph of the overall instrumental setup showing the optics in the sample area and the accessory electronic equipment as described in the text.

ference spectrum. For the measurement of FT-VCD we use two such calibrations, one for the intensity scale and the other for the phase correction, both of which are based on the combined use of a birefringent plate followed by a polarizer at the sample position. In the case of intensity calibration (Nafie *et al.*, 1976; Cheng *et al.*, 1975), the birefringent plate is a multiple quarter-wave plate in the spectral region of interest, each wavelength experiencing a large number of full-wave retardations. For the phase correction, only a small degree of retardation is present so that the entire spectral region is in the same quarter-wave cycle, namely the first such cycle. We shall first consider the intensity calibration.

In Fig. 7 we illustrate four possible orientations of the birefringent plate and second polarizer. In each of the four cases the light from the interferometer is first vertically polarized and the principal modulator stress axis is oriented at 45° with respect to the first polarizer. The fast axis of the birefringent plate can then be either parallel or perpendicular to the modulator axis and the second polarizer may be either parallel or perpendicular to the first one. At a wavelength point for which the multiple quarter-wave plate corresponds exactly to an odd number of quarter-wave retardations, the wave plate will convert alternating left and right circularly polarized light from the modulator to light that is alternating between vertical and

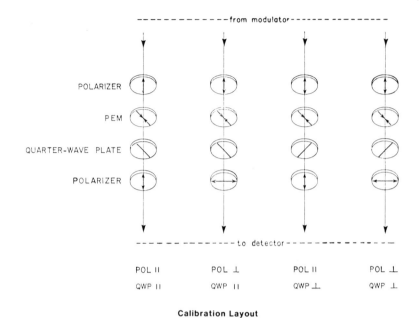

Fig. 7. Optical arrangement for the four possible calibration experiments using a birefringent plate and a second polarizer.

horizontal linearly polarized light. The second polarizer then alternately passes the full intensity of the beam or virtually none of the beam at the modulation frequency. This corresponds to a pseudo-CD signal that is of the same magnitude as the overall instrumental transmission. As the wavelength is scanned from one odd-numbered quarter-wave retardation to the next, the pseudo-CD signal passes through zero and changes to the opposite sign. The normalized expression for this calibration spectrum is (Nafie *et al.*, 1976)

$$\frac{I_{ac}(\bar{\nu})}{I_{dc}(\bar{\nu})} = \frac{\pm 2 J_1[\alpha_0(\bar{\nu})] \, \sin[\alpha_B(\bar{\nu})]}{1 \pm J_0[\alpha_0(\bar{\nu})] \, \cos[\alpha_B(\bar{\nu})]} \tag{29}$$

where $\alpha_B(\bar{\nu})$ is the retardation angle of the birefringent plate at the frequency $\bar{\nu}$. The upper and lower sign choices correspond to the parallel and crossed positions of the two polarizers, respectively. In addition, the sign of the numerator changes upon rotation of the birefringent plate by 90° since $\alpha_B(\bar{\nu})$ changes to $-\alpha_B(\bar{\nu})$.

Since the differential signals from this calibration are very large, it is not difficult to directly record the differential interferogram as it emerges from the lock-in amplifier. In Fig. 8 we show both the ordinary transmission interferogram and the VCD calibration interferogram. It is interesting to note that whereas the transmission interferogram has a large maximum at the point of zero retardation, the calibration interferogram does not. This is because the calibration spectrum consists of nearly equal amounts of positive and negative signal intensity which cancel when they are summed at the zero retardation point.

When the calibration measurement is carried out with a Fourier transform spectrometer, the normalized expression describing the four curves resulting from the experimental arrangements shown in Fig. 7 is given by

$$\frac{FT[V_{ac}(\delta)]}{FT[V_{dc}(\delta)]} = \frac{\pm 2J_1[\alpha_0(\bar{\nu})] \exp[-(2V\bar{\nu})\tau] \sin[\alpha_B(\bar{\nu})]}{1 \pm J_0[\alpha_0(\bar{\nu})] \cos[\alpha_B(\bar{\nu})]} \tag{30}$$

which differs from Eq. (29) by only the exponential factor due to the lock-in amplifier. These four curves are shown in Fig. 9 for the region between 7000 and 1850 cm^{-1}, where the low-frequency limit corresponds to the cutoff of the InSb detector. The strong effect of the exponential factor $\exp[-(2V\bar{\nu})\tau]$ can be seen by the decrease in the magnitude of the extremes with increasing frequency $\bar{\nu}$. This clearly shows that lower-frequency VCD signals will be favored relative to higher-frequency VCD signals.

The curves that go through the positive–negative cycles with the same sign are related by rotation of the birefringent plate and the second polarizer by 90°. These pairs possess the same numerator but opposite signs for J_0 in the denominator of Eq. (30). At the crossing points of these curves above and below zero, they are equal and, therefore, the J_0 term at

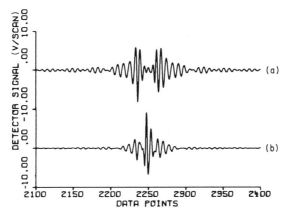

Fig. 8. The VCD calibration interferogram (a) and the transmission interferogram (b) for one of the calibration curves shown in the Fig. 9, from a Nicolet 7199 FT-IR.

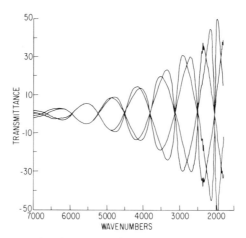

Fig. 9.   Four VCD calibration curves originating from the four optical arrangements in Fig. 7 and described by Eq. (30) in the text. Reprinted with permission from Nafie *et al.* (1979), *J. Am. Chem. Soc.* **101**, 496. Copyright 1979 American Chemical Society.

those points is zero. This occurs only when $\cos[\alpha_B]$ is zero, whereupon $\sin[\alpha_B]$ is ± 1. Therefore, at the crossing points, we have

$$FT[V_{ac}(\delta)]/FT[V_{dc}(\delta)] = \pm 2J_1[\alpha_0(\bar{\nu})] \exp[-(2V\bar{\nu})\tau] \tag{31}$$

This equation, as a function of $\bar{\nu}$, then provides a complete intensity calibration for the VCD intensities described by Eq. (22) since division of Eq. (22) by the calibration curve of Eq. (31) yields the isolated factor (1.1513) $\Delta A_{LR}(\bar{\nu})$.

   In order to carry out the calibration measurement that leads to a proper determination of the phase correction for these differential measurements, we substitute a mechanically stressed piece of ZnSe, as depicted in Fig. 10, in place of the previous birefringent plate. By applying only a small amount of stress, angle α_B is kept small for all wavelengths. Therefore, all intenities in the spectrum will have the same sign, which can be made positive for convenience. With a completely positive spectrum, the phase correction routines can perform normally and a precise phase correction for the complete optical and electronic pathway associated with the VCD measurement can be determined.

D.   Results

   Once the instrument has been calibrated, VCD measurements can be carried out directly by placing only an optically active sample after the modulator, as was shown in Fig. 4. Typical VCD signal intensities are

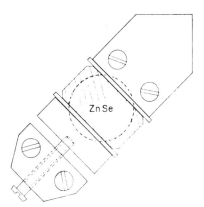

Fig. 10. Diagram of the stressed ZnSe element of a monotonic retardation plate used to obtain a monosignate calibration signal for an accurate determination of the phase correction. The dotted circle represents a hole in the plate on which the vise and the ZnSe element (hatched area) are mounted.

four to five orders of magnitude smaller than the calibration curves in Fig. 9. Consequently, a considerable amount of gain must be applied to the FT-VCD setup in order to observe any effects. This can either be applied at the lock-in amplifier or at the preamplifier stages of the instrument. Adjustments of the final display can be made at the CRT before plotting the results.

1. Initial Experiments

In this section we shall describe a set of early VCD results for the molecule (+)-camphor with a Nicolet 7199 FT-IR spectrometer (Nafie *et al.*, 1979). The designation (+) indicates that the camphor rotates plane polarized through a positive angle in the visible region of the spectrum. The results of VCD measurements in the CH stretching region of 0.5 M (+)-camphor solutions in CCl_4 are given in Fig. 11 along with the molecular structure of camphor. The upper curve in Fig. 11a represents the VCD of this sample recorded by the dispersive VCD instrument at Syracuse University (Diem *et al.*, 1978) and the smooth curve below it is the corresponding transmission, which shows the absorption spectrum in this region. The VCD intensity scale in units of absorbance is shown to the right. The corresponding FT-VCD spectra are shown in Fig. 11b for comparison. A measure of the VCD intensity can be obtained by forming the ratio of $\Delta A/A$, where ΔA is taken as the intensity between the major positive and negative peaks located near 2950 cm^{-1} and A the absorption maximum at 2960 cm^{-1}. The dispersive results yield $\Delta A/A = 8.2 \times$

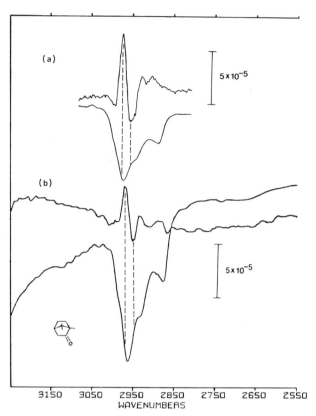

Fig. 11. (a) Dispersive VCD and transmission spectra for 0.5 M (+)-camphor in CCl_4 solution. The scale of VCD intensity is shown in terms of ΔA. The VCD was recorded using a lock-in amplifier with a 10-sec, 18-dB time constant over a period of 66 min, with a resolution of ~ 14 cm^{-1}. A large-area InSb detector (144 mm^2) and a 50 kHz ZnSe modulator were used. (b) FT-VCD of the same sample using the same InSb detector, ZnSe modulator, and lock-in amplifier. The spectrum was obtained from 4096 interferometric scans requiring 160 min with a mirror velocity of 0.145 cm sec^{-1}, an optical filter passing radiation between 3400 and 2400 cm^{-1}, and ~ 8 cm^{-1} resolution, which includes spectral smoothing. Reprinted with permission from Nafie *et al.* (1979), *J. Am. Chem. Soc.* **101**, 496. Copyright 1979 American Chemical Society.

$10^{-5}/0.77 = 10.6 \times 10^{-5}$, and the FT result is $\Delta A/A = 5.2 \times 10^{-5}/0.76 = 6.8 \times 10^{-5}$, which shows reasonably close agreement considering the small magnitude involved. Another spectrum of this sample (Nafie *et al.*, 1976) gives $\Delta A/A = 8.4 \times 10^{-5}$, which falls between these results. The FT-VCD results were obtained from four blocks of 1024 interferometric scans. The resolution was 2 cm^{-1} with smoothing applied to the final result to yield approximately 8 cm^{-1} resolution. By comparison, the resolution of the dispersive VCD spectrum was set to be approximately 14 cm^{-1} over the region displayed. The time required to obtain the FT-VCD was 160 min as compared to 66 min for the region scanned in the dispersive VCD.

In an overall sense, the FT-VCD spectrum shows comparable, but somewhat less favorable, results relative to the dispersive spectrum. At this point, it is clear that genuine VCD signals are being measured but a number of aspects need to be improved before the performance is raised to the level of the dispersive spectrum for a narrow spectral region. Note, however, that the FT-VCD results extend beyond the limits of the wavenumber display in Fig. 11. Unfortunately, there are no additional major absorption bands to investigate in this region. If there were, they would have appeared simultaneously with present FT-VCD results whereas additional time would have been required to scan the corresponding dispersive spectrum.

A further verification of our FT-VCD results is given in Fig. 12 where the previous FT-VCD spectrum for (+)-camphor is compared to that for

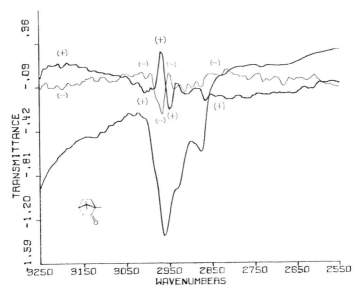

Fig. 12. FT-VCD and transmission spectra of (+) and (−)-camphor under the same conditions described in Fig. 11b except that 1024 scans requiring 40 min were used to obtain the FT-VCD of the (−)-camphor. Reprinted with permission from Nafie *et al.* (1979), *J. Am. Chem. Soc.* **101**, 496. Copyright 1979 American Chemical Society.

the mirror image compound, $(-)$-camphor. Only one block of 1024 scans was recorded for $(-)$-camphor, which results in a somewhat higher noise level. The two major peaks in the spectrum are of approximately the same intensity but with opposite signs, as expected for the CD of mirror image pairs of molecules. The smaller features are also opposite but their detailed shapes are distorted by the noise level.

An interfering problem encountered in our FT-VCD measurements was a distorted VCD base line, which depended, in part, on the transmission spectrum. The transmission dependence could be removed by subtracting a small but constant fraction of the transmission spectrum from each of our VCD results. These interfering effects originate in imperfections in the optical path resulting in unwanted birefringence that can be detected by polarization sensitivity in the InSb detector. We were able to reduce these effects by using a large-area InSb detector (144 mm^2, Spectronics, Inc.) and imaging the ir beam over the most of the detector surface to allow averaging of local regions of differing polarization sensitivity. As a consequence, we avoided tight focusing of the beam onto a small detector element (1 mm^2) in our initial VCD experiments.

2. Continued Experiments

In order to improve the performance of the FT-VCD setup, two major instrument changes were carried out. A ZnSe modulator operating at a frequency near 100 kHz replaced a 50-kHz model used in the initial experiments, and a new "sandwich" infrared photodetector comprised of an InSb element (3 mm in diameter) was mounted above a HgCdTe (MCT) element (1 mm in diameter). The ZnSe modulator, due to its higher operating frequency, permits a more efficient electronic filtering of the VCD interferogram from the transmission interferogram. The "sandwich" detector provides continuous coverage from ~ 900 to 10,000 cm^{-1}. Above 1850 cm^{-1}, the InSb element is in operation. Its element size was chosen to be considerably smaller than the large-area InSb detector used in the initial experiments (with an attendant factor of ~ 4 reduction in the detector noise level) and yet large enough both to provide averaging of polarization sensitivity over the detector surface and to prevent detector saturation of the InSb element when the intensity from the full aperature of the instrument is focused uniformly over its surface. Below 1850 cm^{-1}, the InSb element becomes nearly transparent and acts as a 77°K low-wavelength-pass optical filter, thereby greatly decreasing the noise of the MCT detector and increasing the saturation threshold by blocking radiation above 1850 cm^{-1}. In addition, a unified grounding and cable network has eliminated electronic interference encountered in the initial experiments, and a provision for switching between the VCD interferogram signal and the transmission interferogram signal has been installed.

Figure 13 shows the new FT-VCD result for a 0.5 M solution of (+)-camphor in CCl_4. Only 22 min were required to obtain this result, which shows a superior signal-to-noise ratio compared to the corresponding FT-VCD spectrum in Fig. 11b (which required 160 min of signal averaging). In addition, no spectral smoothing was applied in the more recent spectrum. The only drawback of the new spectrum is a more serious distortion of the VCD peaks arising from an irregular (nonflat) base line. This distortion, which appears even for a nonoptically active absorption, has been traced back to hydrocarbon impurities (presumably due to vacuum pump oil) embedded within the 100-kHz ZnSe modulator crystal. Some of this impurity absorption was due to surface contamination and was easily removed by organic solvents. However, a residual absorption in the CH stretching region ($\sim 1\%$ intensity reduction) could not be removed. As mentioned earlier (Section V.D.1), all of our FT-VCD spectra initially exhibit a distortion that depends on the sample absorption. Although this distortion is a function of the optical alignment, it is a constant factor for a given alignment and can be removed by subtraction. However, the distortion encountered in the CH stretching region with the new modulator appears to be a second-order effect and is not removable by simple subtraction procedures.

In order to test the FT-VCD setup in a different spectral region (where the ZnSe modulator shows no impurity absorption), a 0.01 M solution in CCl_4 of S(+)-1-carboxy-1-ethyl-3-phenylallene was investigated in its $C{=}C{=}C$ antisymmetric stretching made near 1950 cm^{-1}. The absorption and VCD results for this mode are shown in Fig. 14, where the FT measurements are compared to the corresponding dispersive spectra obtained at the University of Minnesota (Abatte *et al.*, 1980). Only 18 min were

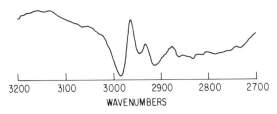

3200   3100   3000   2900   2800   2700
WAVENUMBERS

Fig. 13. FT-VCD spectrum for (+)-camphor in CCl_4 using a 100-kHz ZnSe modulator and a 3-mm-diameter InSb detector. The mirror velocity was 0.133 cm sec^{-1} and the data (2048 points per scan) were collected with 1024 scans requiring less than 22 min of acquisition time (1302.58 sec) at a resolution of 6 cm^{-1}, without spectral soothing. The same optical filter used in the earlier measurements was employed.

required to obtain the FT-VCD, which shows virtually a noise-free result, whereas considerable noise is present in the conventional VCD spectrum, which, in addition, required a longer period of time (~ 1 hr) to obtain. A direct comparison of the FT-IR and dispersive VCD spectra is given in Fig. 14c, where the two results are plotted on the same $\Delta A/A$ scale. The close agreement between the two sets of spectra indicates that the FT-VCD result is accurate and substantially free of interfering artifacts and that base line distortion is not a problem here.

As a further test of the capability of the new FT-VCD setup, we measured the carbonyl (C=O) stretching vibration at 1760 cm^{-1} in (+)-3-bromocamphor. Here we used the MCT detector for the first time and although the noise level is higher (due to a lower D^* for these detectors), the noise was considerably reduced from normal MCT levels due to the cold-shielding effect of the InSb element. The results of FT-VCD as well as dispersive VCD measurements carried out in our laboratory at Syracuse for this vibrational band are given in Fig. 15. Again the FT-VCD shows virtually no noise, whereas the dispersive results contain rather large noise contributions. The dispersive result was obtained from a 1-hr scan, whereas the FT-VCD resulted from averaging nine individual mea-

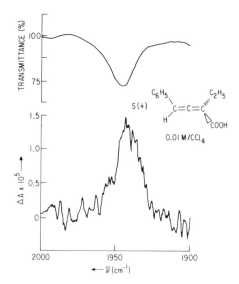

Fig. 14a. VCD and absorption spectra obtained using dispersive methods for the C=C=C antisymmetric stretching vibration in a 0.01 M solution of S(+)-1-carboxy-1-ethyl-3-phenylallene in CCl$_4$. The dispersive VCD was measured at the University of Minnesota using a resolution of 12 cm^{-1} with a scan time of approximately 1 hr and a CaF$_2$ photoelastic modulator.

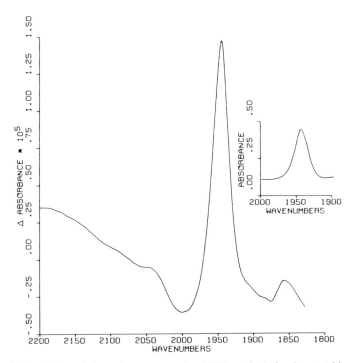

Fig. 14b. VCD and absorption spectra using FT-IR methods for the stretching of Fig. 14a. The FT-VCD result was obtained after 18 min with 1024 interferometric scans, a mirror velocity of 0.133 cm sec^{-1}, a resolution of 6 cm^{-1} without smoothing, and no optical band pass filter.

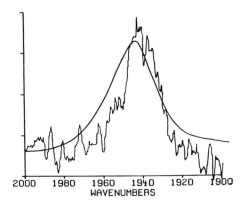

Fig. 14c. The VCD in Figs. 14a and b replotted using the same $\Delta A/A$ scale, which gives a direct comparison of these two results.

surements requiring approximately 25 min each. To gain an appreciation for the degree of consistency of these nine individual spectra, we show them in Fig. 16. Each was obtained in less than half the time required to observe the dispersive VCD spectrum, yet they possess considerably less noise and a high degree of reproducibility. Another significant point to

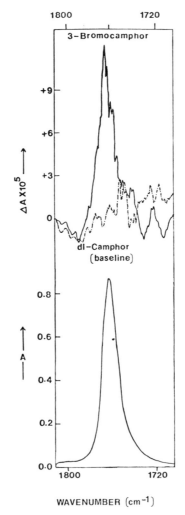

WAVENUMBER (cm^{-1})

Fig. 15a. VCD and absorption spectra obtained using dispersive methods for the carbonyl (C=O) stretching vibration in a 0.5 M solution of (+)-3-bromocamphor in CCl$_4$ solution. The dispersive spectra were measured at Syracuse University using 1 × 5 mm MCT detector, a CaF$_2$ photoelastic modulator, and 60 min of scan time at a resolution of 10 cm^{-1}. A VCD base line was obtained by using a solution of *dl*-camphor.

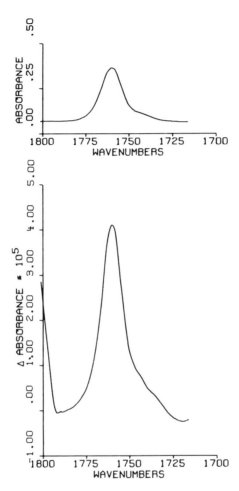

Fig. 15b.   VCD and absorption spectra obtained using FT-IR methods for the stretching of Fig. 15a. The FT-VCD spectrum using a cold-shielded 1-mm-diameter MCT detector,
and a 100-kHz ZnSe modulator was obtained using nine blocks of 1024 scans each, with a
total acquisition time of 3.8 hr, a scan speed of 0.133 cm sec^{-1}, and a resolution of 6 cm^{-1}
with no smoothing. No optical band-pass filter was employed.

bear in mind while comparing the Fourier to the dispersive spectra is that
only a water-cooled glower source (~1200°C) was used for FT-VCD,
whereas either a tungsten–halogen lamp (~3000°C) or a xenon lamp
(~6000°C) was used for the dispersive VCD. To a first approximation, the
higher-temperature sources provide more radiation as the fourth power of
the absolute temperature, thus making it doubtful that conventional VCD
could be observed at all in most cases without the brighter sources. On

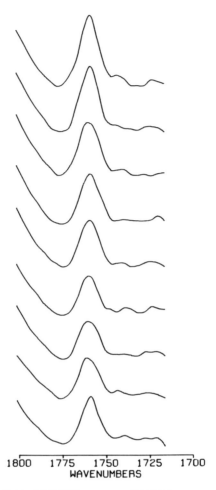

1800 1775 1750 1725 1700
WAVENUMBERS

Fig. 16. Nine individual FT-VCD spectra of (+)-3-bromocamphor generated from 1024 interferometric scans (25 min each) showing the degree of reproducibility of the spectra used to obtain the FT-VCD result in Fig. 15b.

the other hand, these bright sources have inconvenient lower-wavelength cutoffs due to their optical enclosures (4.5 μm, quartz, for tungsten lamp and 6.5 μm, sapphire, for the xenon lamp). In addition, they can saturate infrared photodetectors in an FT-IR spectrometer and need to be filtered to a narrower optical range to avoid this problem (thus decreasing the Fourier advantage). As a final point, we note that although optical filtering was helpful in reducing noise and avoiding detector and ADC saturation in the early FT-VCD measurements with InSb near 3000 cm^{-1}, it was not necessary in the more recent work near 2000 cm^{-1} for the allene spectra or the carbonyl measurements with the MCT detector.

E. Noise Analysis

We shall conclude our consideration of FT-VCD by first considering the various sources of noise that may appear in the spectra. We shall then go on to demonstrate with base line noise spectra that our recent FT-VCD measurements are not limited by noise sources other than the detector itself. The implications of this result for the eventual routine measurement of FT-VCD spectra over wide spectra ranges will then be discussed.

The noise level of our FT-VCD spectra is the product of the inherent noise of the instrument detection and the noise contribution from the lock-in detection. The sensitivity of an FT-IR spectrometer has been measured and shown to agree with theoretical predictions by Mattson (1978). His calculations assume that all information can be passed by the instrument's 15-bit analog-to-digital converter (ADC). As he points out, this assumption is valid for his experiment, but other, more-sensitive detectors (such as are used for VCD) have larger dynamic ranges. These detectors are thus able to send signals that are too large to the ADC. This leads to the loss of small spectral detail and the appearance of digital noise resulting from the truncation (at 15 bits) of these large intensities. Double modulation can be used to bypass this limitation because the signal remaining after analog lock-in detection of the detector signal contains only the interferogram corresponding to the difference spectrum. Again, this difference interferogram has a far lower dynamic range because all of the usual spectral energy information has been subtracted out.

A price is paid for this advantage, however. First, the optical and electronic phase information is no longer in the interferogram. As explained earlier, this information can be obtained from a reference spectrum, so this is not really a problem if the interferometer is stable enough. Second, the modulation–demodulation process degrades the signal. Double modulation is worthwhile only when there is a sufficiently large ratio between the modulation frequency and the signal frequencies (modulation index). In the practical demodulation system, the time-varying modulation frequency is truncated in time by the filters. Sharp-cut filtering introduces phase errors into the modulated (ac) signal, while less severe filtering leads to amplitude errors and incomplete elimination of the unmodulated (dc) interferogram. The mathematics of filter transfer functions has been worked out in detail by Gold and Rader (1969), and Harris (1978) has calculated and compared a variety of transfer functions. Additional problems can be caused when the modulation frequency is asynchronous with the

FT-IR sampling frequency (as present here) since the difference frequency and its harmonics then exist as a noise contribution to the signal.

In Fig. 17 we show a set of five base line noise spectra on the same scale of sensitivity, and although they represent different measurement conditions and procedures, they can be compared directly to one another without scale adjustment. The spectral region here is from 900 to 1300 cm^{-1} using the cold-shielded MCT detector. This region was chosen because it displays no base line distortion nor any significant atmospheric absorption bands. These measurements were also repeated between 1400 and 1800 cm^{-1} under the same instrumental conditions where interfering absorption does occur (the instrument was purposely left unpurged to test for the effect of this absorption), and these results are given in Fig. 18.

Figures 17a–c represent noise measurements of ordinary FT-IR transmission spectra. They are normalized twice in the sense that, first, two consecutive single-beam spectra are subtracted to remove spectral features, leaving only noise. This difference spectrum is then normalized to an overall transmission so that noise spectra independent of the light level or differing light levels in different spectral regions are obtained. We refer to the final noise spectrum as a subtraction product and pairs of repeated subtraction products are displayed for each experiment except the last one in Fig. 17e. The clear-beam subtraction products in Fig. 17a contain a slightly larger noise level relative to the completely blocked beam spectra. The clear beam tests the ability of the ADC to measure large intensities accurately (its most significant bits), whereas the blocked beam tests its accuracy at zero intensity (its least significant bits). In order to ensure that digitization noise is not present, the preamplifier gain was increased by a factor of 128 in the blocked-beam experiment and little, if any, change in the noise level occurred. Thus the noise seen in the three transmission experiments is due almost entirely to the MCT detector. Figure 17d contains subtraction products associated with an FT-VCD base line. The raw VCD curves were first subtracted, then ratioed to the clear-beam transmission, and scaled in accordance with the quarter-wave-plate calibration procedure described previously. The low dynamic range of the VCD spectra precludes the presence of digitization noise. However, noise due to polarization modulation followed by lock-in demodulation (synchronization noise) may be present. Although some increase in the noise is present in the VCD subtraction products, this additional contribution is not dominant, as can be deduced by comparison with Figs. 17a–c. In Fig. 17e, sixteen times as many scans are used to obtain the spectra for the subtraction product and this results, as expected, in a fourfold reduction of the noise level. We conclude that the noise in our FT-VCD spectra using an MCT detector arises principally from the detector itself.

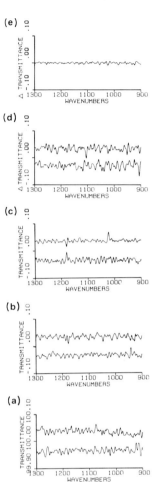

Fig. 17. Noise levels in transmission and VCD FT-IR using an InSb cold-shielded MCT (850 cm^{-1} bandgap) detector. The ZnSe PEM was positioned in the beam for all measurements. Each trace represents a 16-scan ($\sim$ 10-sec) run subtracted from a similar 16-scan run, so the noise level of the trace is $\sqrt{2}$ higher than the noise level in an individual run. Dual-displaced spectra represent repeated experiments. (a) Clear-beam spectral subtraction products. The individual spectra were ratioed to a single-beam background (for scale) and converted to absorbance before subtraction. Then the subtraction product was reconverted to a transmittance. (b) Blocked-beam subtraction products. The individual spectra (each ratioed to a single clear-beam background) were subtracted in transmittance. (c) Same as (b) but with electronic gain increased 128 times to eliminate the digitizer dynamic range as a possible limitation. (d) Clear-beam VCD subtraction products. Individual VCD spectra (each ratioed to a clear-beam spectrum and scaled according to a quarter-wave-plate calibration plot) were subtracted in "transmittance." (e) Same as (d) but the subtraction products result from two 256-scan runs.

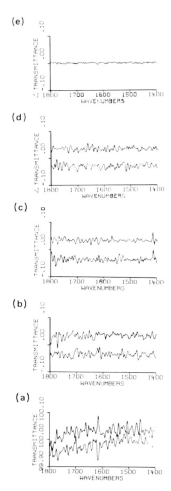

Fig. 18. Same as Fig. 17 but showing a spectral region with considerable water vapor interference (the spectrometer beam path was incompletely purged during these experiments). Although the clear-beam transmission noise is somewhat increased, there is no significant effect on the VCD noise level.

In Fig. 18, the additional complication of narrow atmospheric absorption lines creates increased digitization noise in the clear-beam transmission measurement due to digital inaccuracies when two such consecutive spectra are subtracted and normalized. However, blocking the beam removes this source of noise and increasing the electronic gain again shows the absence of digitization noise in the least significant bits of the ADC. The VCD subtraction products (clear beam) show none of the digitization difficulties present in Fig. 18a and little, if any, increase (compared to Figs. 18b and c) due to synchronization noise. Again, increasing the number of scans improves the VCD noise spectrum as expected.

From all of these noise measurements we can conclude that in the present setup DM-FT-VCD is not limited by the ADC dynamic range or by preamplifier or lock-in amplifier noise contributions; only detector noise makes a major contribution. This conclusion holds for the $\sim 900-1850$ cm^{-1} region where the MCT detector is in operation. Above 1850 cm^{-1}, the InSb detector possesses a noise contribution that is nearly an order of magnitude smaller than the MCT detector. Although a detailed noise analysis has not been carried out, our VCD of the substituted allene near 2000 cm^{-1} indicates that synchronization is beginning to enter as a noise-limiting factor. Furthermore, more significant digitization noise can now be expected for the clear-beam-transmission subtraction products in this region. Similar conclusions are also valid in the C–H stretching region. However, here the base line distortion problem imposes an unsolved optical limitation on the acquisition of accurate VCD spectra. One way around this problem is to use a CaF_2 modulator for the higher vibrational frequencies (down to ~ 1300 cm^{-1} if desired) and a ZnSe modulator at lower frequencies.

Given the low levels of noise contributions from either the ADC or the modulation–demodulation optoelectronics, the prospects for measuring VCD with nearly the full realization of the Fourier advantage relative to a dispersive VCD spectrometer appear likely. No major obstacle is apparent at this time and the way seems clear for the eventual reduction of FT-VCD to routine measurement. This optimistic prognosis must now await further testing, evaluation, and development.

VI. CONCLUSIONS

Clearly, there exists a wide range of potential applications of double modulation Fourier transform spectroscopy. As a general new approach to infrared spectral measurement, it has only just begun to be investigated. The fact that this method has proved successful in the measurement of vibrational circular dichroism, which, until a few years ago, was undiscovered due to its small magnitude, indicates that this approach is capable of high sensitivity and should be extended in other directions. Particularly intriguing is the possibility of studying transient phenomena in chemical dynamics, such as reaction intermediates where the full infrared spectrum can be used as the reaction probe. Significant advances in studying transient species in vibrational spectroscopy have been recently discussed by Durana and Mantz (1979) for FT-IR and by Woodruff and

Farquarson (1978) for pulsed-resonance Raman scattering. In both of these reviews, the use of vibrational spectral multiplexing is the key ingredient that provides such an enormous advantage over earlier kinetic studies.

In terms of further advances in VCD, the most significant area of development appears to be the extension of the spectral range into the mid-infrared region where the advantages of DM-FT grow with respect to conventional dispersive methods. These advantages result primarily from the lower Fourier frequencies associated with lower vibrational frequencies that, in turn, result in a larger modulation index (Section V.E). Due to the fact that the mid-infrared region is vitally important in the study of molecular vibrations and that VCD measurements in this region are difficult and just beginning to appear (Stephens and Clark, 1979), the investigation of FT-VCD in this region will be a significant next step.

REFERENCES

Abatte, S., Havel, H. A., Laux, L., Moscowitz, A., and Overend, J. (1980). Unpublished results.

Adamowicz, R., and Fishman, E. (1972). *Spectrochim. Acta* **A28**, 889.

Barron, L. D. (1978). *In* "Advances in Infrared and Raman Spectroscopy" (R. J. H. Clark and R. E. Hester, eds.), Vol. IV, p. 271. Heyden, London.

Barron, L. D. (1979). *In* "Optical Activity and Chiral Discrimination" (S. F. Mason, ed.), p. 219. Reidel, Dordrecht, Holland.

Bell, R. J. (1972). "Introductory Fourier Transform Spectroscopy." Academic Press, New York.

Boccara, A. C., Duran, J., Briat, B., and Stephens, P. J. (1973). *Chem. Phys. Lett.* **19**, 187.

Cheng, J. C., Nafie, L. A., and Stephens, P. J. (1975). *J. Opt. Soc. Am.* **65**, 1031.

Cheng, J. C., Nafie, L. A., Allen, S. D., and Braunstein, A. I. (1976). *Appl. Opt.* **15**, 1960.

Connes, J., and Connes, P. (1966). *J. Opt. Soc. Am.* **56**, 896.

D'Espisito, L., and Koenig, J. L. (1978). *In* "Fourier Transform Infrared Spectroscopy" (J. R. Ferraro and L. J. Basile, eds.), Vol. 1, p. 61. Academic Press, New York.

Diem, M., Gotkin, P. J., Kupfer, J. M., Tindall, A. G., and Nafie, L. A. (1977). *J. Am. Chem. Soc.* **99**, 8103.

Diem, M., Gotkin, P. J., Kupfer, J. M., and Nafie, L. A. (1978). *J. Am. Chem. Soc.* **100**, 5644.

Dignam, M. J., and Baker, M. D. (1980). Unpublished results.

Durana, J. F., and Mantz, A. W. (1979). *In* "Fourier Transform Infrared Spectroscopy." (J. R. Ferraro and L. J. Basile, eds.), Vol. 2, p. 1. Academic Press, New York.

Fellgett, P. B. (1958). *J. Phys. Radium* **19**, 187.

Forman, M. L. (1966). *J. Opt. Soc. Am.* **56**, 978.

Forman, M. L., Steel, W. H., and Vanasse, G. A. (1966). *J. Opt. Soc. Am.* **56**, 59.

Gendreau, R. M., and Griffiths, P. R. (1976). *Anal. Chem.* **48**, 1910.

Gold, B., and Rader, C. M. (1969). "Digital Processing of Signals." McGraw-Hill, New York.

Griffiths, P. R. (1975). "Chemical Infrared Fourier Transform Spectroscopy." Wiley, New York.

Griffiths, P. R. (1977). *Appl. Spectrosc.* **31**, 497.

Griffiths, P. R. (1978a). *In* "Fourier Transform Infrared Spectroscopy" (J. R. Ferraro and L. J. Basile, eds.), Vol. 1, p. 143. Academic Press, New York.

Griffiths, P. R. (1978b). *In* "Transform Techniques in Chemistry" (P. R. Griffiths, ed.), pp. 109 and 141. Plenum, New York.

Griffiths, P. R. (1978c). Private communication.

Grosjean, M., and Legrand, M. (1960). *C. R. Acad. Sci. (Paris)* **251**, 2150.

Harris, F. J. (1978). *Proc. IEEE* **66**, 51.

Hirschfeld, T. (1975). *Appl. Spectrosc.* **29**, 524.

Hirschfeld, T. (1976). *Appl. Spectrosc.* **30**, 550.

Hirschfeld, T. (1979). *In* "Fourier Transform Infrared Spectroscopy" (J. R. Ferraro and L. J. Basile, eds.), Vol. 2, p. 193. Academic Press, New York.

Hofrichter, J., and Eaton, W. A. (1976). *Annu. Rev. Biophys. Bioeng.* **5**, 511.

Holzwarth, G., Hsu, E. C., Mosher, H. S., Faulkner, T. R., and Moscowitz, A. (1974). *J. Am. Chem. Soc.* **96**, 251.

Jacquinot, P. (1960). *Rep. Prog. Phys.* **13**, 267.

Kemp, J. C. (1969). *J. Opt. Soc. Am.* **59**, 950.

Koenig, J. L. (1975). *Appl. Spectrosc.* **29**, 293.

Kuehl, D., and Griffiths, P. R. (1978). *Anal. Chem.* **50**, 418.

Kusan, T., and Holzwarth, G. (1976). *Biochemistry* **15**, 3352.

Martin, D. E., and Puplett, E. (1969). *Infra. Phys.* **10**, 105.

Mattson, D. R. (1978). *Appl. Spectrosc.* **32**, 335.

Mertz, L. (1967). *Infra. Phys.* **7**, 17.

Nafie, L. A., Diem, M. (1979a). *Appl. Spectrosc.* **33**, 130.

Nafie, L. A., and Diem, M. (1979b). *Acc. Chem. Res.* **12**, 296.

Nafie, L. A., and Vidrine, D. W. (1979). *Proc. Soc. Photo. Inst. Eng.* **191**, 56.

Nafie, L. A., Cheng, J. C., and Stephens, P. J. (1975). *J. Am. Chem. Soc.* **97**, 3842.

Nafie, L. A., Keiderling, T. A., and Stephens, P. J. (1976). *J. Am. Chem. Soc.* **98**, 2715.

Nafie, L. A., Diem, M., and Vidrine, D. W. (1979). *J. Am. Chem. Soc.* **101**, 496.

Painter, P. C., and Koenig, J. L. (1977). *J. Polym. Sci. Part A-2* **15**, 1885.

Piseri, L., Cabassi, F., and Masseti, G. (1975). *Chem. Phys. Lett.* **33**, 378.

Russel, M. I., Billardon, M., and Badoz, J. P. (1972). *Appl. Opt.* **11**, 2375.

Stephens, P. J., and Clark, R. (1979). *In* "Optical Activity and Chiral Discrimination" (S. F. Mason, ed.), p. 263. Reidel, Dordrecht, Holland.

Strassburger, J., and Smith, I. T. (1979). *Appl. Spectrosc.* **33**, 283.

Velluz, L., Grosjean, M., and Legrand, M. (1965). "Optical Circular Dichroism." Academic Press, New York.

Vidrine, D. W., and Nafie, L. A. (1979). Unpublished results.

Wada, A. (1972). *Appl. Spectrosc. Rev.* **6**, 1.

Woodruff, W. H., and Farguarson, S. (1978). *In* "New Applications of Lasers in Chemistry" (G. M. Hieftje, ed.). American Chemical Socity, Washington, D.C.

4
PHOTOACOUSTIC FOURIER TRANSFORM INFRARED SPECTROSCOPY OF SOLIDS AND LIQUIDS

D. Warren Vidrine

Nicolet Instrument Corporation
Madison, Wisconsin

I. INTRODUCTION

Curiously, 1980 marks the centennial year, not only of the discovery of the photoacoustic (pa) effect, but also of the invention of the interferometer. In this chapter, the combination of these important spectroscopic tools, Fourier transform infrared photoacoustic spectroscopy, FT-IR/PAS (also called PA-FT-IR), will be introduced. This measurement technique allows ir spectra of solid samples to be obtained essentially without preparation. Infrared photometers using pa detection have

125

Copyright © 1982 by Academic Press, Inc.
All rights of reproduction in any form reserved.
ISBN 0-12-254103-0

been used for years (Veingerov, 1938) and pa detection in modern laser photometers is commonplace. However, individual laser photometers characteristically are not capable of scanning more than a small fraction of the mid-ir spectrum, and dispersive ir measurements (e.g., see Low and Parodi, 1980) are quite difficult because of several factors, including the low intensity of thermal sources and the relative insensitivity of the air microphone cell as an ir detector (especially with condensed-phase samples). The advent of FT-IR/PAS allows the information-rich mid-region to be routinely measured using pa detection. Photoacoustic spectroscopy (pas) has already been shown to have valuable advantages over other special measurement techniques in the visible region, and the major limitation of pas in the uv–visible (signal saturation) is not a serious problem in the infrared. Present applications include the measurement of solid samples without the need for sample preparation, the avoidance of the Christiansen effect in band-shape measurements, the investigation of the aging chemistry of freshly cleaved surfaces, and the quantitative analysis of a surface coating on a rough absorbing substrate (clay). The use of FT-IR/PAS should be considered whenever physical sample preparation is difficult or when the occurrence of an artifact of preparation is suspected. A problem arises if the grinding, dissolution, melting, adulteration, deformation, or polishing required for other measurement techniques (transmission, reflectance, emission) occur or if the preparation introduces an undefined or obscuring artifact into the measured spectrum.

II. HISTORY

A. G. Bell is certainly best known for his work on electroacoustic transducers, but he is also the discover of the photoacoustic pa effect (Bell, 1880, 1881). Michelson, a contemporary and acquaintance, developed the interferometer as an extension of his work on the speed of light (Michelson, 1881). In fact, Bell arranged the research grant (Michelson's first) that funded the development of the interferometer. These two great experimentalists probably never foresaw the mating of their two inventions (although Bell prophesized that the most important application of his "spectrophone" would be in the ir region), but further developments had to await the invention of suitable transducers.

The next development was the intermittent use of tuned pa cells in photometry, culminating in the now widespread pa detection of gases in laser photometers. In 1973, Rosencwaig developed the untuned air-microphone pa cell, which made possible the first commercial pa spectrometer. At this point, both the untuned pa cell and the FT-IR spectrometer were commercial realities and it was now only a matter of time before they were combined. The first interferometric pa spectrum (of methanol vapor) was reported by Busse and Bullemer (1978). Farrow *et al.* (1978) extended the technique to solids, working in the visible range, and the next year, Vidrine (1979) and Rockley (1979) independently obtained the first Fourier transform infrared photoacoustic spectra of solids. Vidrine (1980) further investigated the applications of FT-IR/PAS, presenting examples of routine pa measurements of otherwise difficult-to-prepare samples and comparing results obtained with FT-IR/PAS to those obtained with ATR and diffuse reflectance techniques.

The applications of FT-IR/PAS are now expanding rapidly. Royce's group (Royce *et al.*, 1980; Laufer *et al.*, 1980) is using the technique to obtain spectra of inorganic salts free of Christiansen bands. Rockley and Devlin (1980) are investigating the aging of freshly cleaved coal surfaces.

Lowry *et al.* (1981) are measuring the concentration of pesticide on clay substrates. Nicolet Instrument Corporation introduced the Gilford/Nicolet pa cell accessory for their FT-IR spectrometers at the 1980 Pittsburgh Conference on Analytical Chemistry and Applied Spectroscopy (Atlantic City, New Jersey). Also, EG&G Princeton Applied Research has redesigned their pa cell (which has gas exchange capability) for use with FT-IR. These cells (illustrated in Fig. 1) are designed for placement in the FT-IR sample compartment.

III. THEORY

The qualitative–quantitative theory of the photoacoustic effect has been thoroughly worked out by uv–visible PAS researchers. The applications of this theory to FT-IR practice are particularly simple because the complexities of complete signal saturation are avoided automatically with most common ir chromophores. The only real difference stems from the multiplex nature of FT-IR: in a FT-IR/PA spectrum, the modulation frequency for each spectral element is proportional to the wavenumber of that element. However, any spectrum obtained at a certain mirror velocity can be compared perfectly with others obtained at that same velocity, and this difference is only significant in depth-profiling studies and investigations of fluorescence and phosphorescence.

A. Origin of the Photoacoustic Effect

Light energy absorbed by a confined gas heats it and causes an increase in pressure. Modulated light absorbed by a gas produces a pressure modulation in the gas. A sound is generated at the frequency of the light modulation and can be detected with a microphone. As diagrammed in Fig. 2, the photoacoustic effect in solids involves a thermal transfer from the solid sample to the surrounding air. Energy absorbed by the solid heats its surface, which in turn conductively heats a boundary layer of air next to the solid surface. Modulated light energy absorbed at the solid surface therefore causes intermittent expansion of this boundary layer, producing sound. The major feature that makes this effect valuable for solid-sample spectroscopy is the fact that only the energy absorbed near the sample surface is effective in heating the surface quickly enough to contribute to the photoacoustic signal. This means that an optically thick, totally absorbing sample will give a useful, nonsaturated, pa spectrum as long as the sample is not completely absorbing at the effective thermal diffusion depth u_s. This depth is determined by the thermal conductivity and the

(a)

Fig. 1. Commercially available PA-FT-IR cells: (a) Gilford/Nicolet pa cell; (b) Princeton Applied Research pa cell (version shown is adapted for use with the Nicolet 7199 FT-IR).

(b)

Fig. 1   *(Continued)*

heat capacity of the sample and by the modulation frequency. Kreuzer (1977) has reviewed the physics of the pa effect in gases, and Rosencwaig (1977, 1978; see also Rosencwaig and Gersho, 1976; and McDonald and Wetsel, 1978) has developed a comprehensive theory for the photoacoustic effect as applied to solid samples. The reader is referred to these sources for the detailed qualitative–quantitative theory of photoacoustic spectra. For the purposes of this chapter, two simple approximations are adequate. First, pa spectra are qualitatively and quantitatively similar to transmission spectra, are commonly ratioed to a saturated reference (carbon black), and these ratioed spectra are approximations of inverted transmission spectra (base line is at 0% and saturated peaks approach 100%). Second, sample surface morphology exerts a minimal qualitative effect on the spectra. Chemically identical samples with widely different surfaces (smooth or rough, scattering or nonscattering) have the same spectral features, similar peak intensity ratios, and a characteristically flat base line.

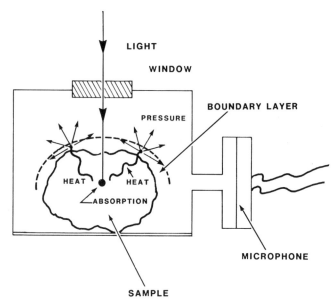

Fig. 2. Idealized pa cell for solid samples.

B. Signal Phase and Amplitude

The photoacoustic signal resulting from light absorption within the sample is reduced in amplitude and delayed in phase because the heat must be conductively transmitted to the sample surface before it can contribute to the acoustic signal. With a practical, optically thick sample, then, overall signal phase is a function of the "average" optical absorption depth (u_s, the inverse of the Napierian or base e volume absorptivity) and therefore varies considerably over a spectrum. In FT-IR/PAS, phase can be ignored completely by calculating the intensity (power, or rms) spectrum. This calculation unfortunately rectifies noise where there is no signal, giving a slightly positive value (equal to the rms noise level) to otherwise zero base line areas. Phase correction, using a symmetrical, double-sized interferogram, allows a true phase-corrected spectrum, without the base line offset artifact, and is the calculational method of choice for most work. The traditional single-sided interferogram (with a few hundred points on the other side for calculation of a low-resolution phase correction) produces significant spectral artifacts because the large pa phase variations across typical (10–20 cm^{-1} wide) spectral bands cannot be phase-corrected properly. Spectra of this type remain similar in overall appearance to the true spectra, and noncritical analytical work can probably still be done with them.

The phase spectrum itself has already proved valuable for some applications. Robin's group have elegantly measured phosphorescence lifetimes and quantum yields in the uv–visible using phase pa spectra, and Robin's review of this and other work should be referred to by interested readers. Phase spectra are easily obtainable in FT-IR by comparison of the single-sided and double-sided transforms. Optical, transducer, and electronic contributions to phase error are also included in these spectra, but they can be eliminated in pa spectra of solid samples by subtracting the phase spectrum of carbon black. Phase pa spectra of solid samples can be used for depth profiling, as will be described later.

IV. OPERATIONAL CHARACTERISTICS

A. Effect of Gas Composition and Pressure

Intuition tells us that air at atmospheric pressure is probably not the best energy transfer medium for a pa cell. Wong (1978) investigated several gases at atmospheric pressure and concluded that helium was the best of those tested. However, the pressure dependence of the pa signal has not been adequately explored, and further improvement of PA-FT-IR sensitivities may be possible by optimizing cell pressure. A recent study using an absorbing gas (Wake and Amer, 1979) serves as an illustration, although the results are not directly applicable to pa spectroscopy of solids.

The further elucidation of the effect of gases and pressures is useful for another reason. The investigation of surface species on catalysts is a promising area for FT-IR/PAS applications, and the pressure and gas composition will often be dictated by the experimental conditions required. Can sensitivity be improved by altering or adding a buffer gas? How can the relative amounts of signal from the sample surface and from an absorbing gas be changed?

B. Effect of Modulation Frequency

Because FT-IR spectrometers modulate different frequencies of light at different audio frequencies, pa cells designed for use with FT-IR spectroscopy must respond to a range of modulation frequencies. The untuned air-microphone cell originally developed by Rosencwaig (1973; Blank and Wakefield, 1979) fulfills this requirement, and similar cells have been used for all FT-IR/PAS work published to date. A characteristic of this cell is the low, broad, Helmholtz resonance at approximately 1 kHz and the inverse relationship of signal strength and modulation frequency as shown in Fig. 3. These cells are more sensitive at lower modulation frequencies, and using a low interferometer mirror velocity places the modulation frequencies corresponding to the mid-ir spectrum in the low audio range, where cell sensitivity is highest. The measurement-parameter optimization for the Blank and Wakefield cell resulted in a mechanical mirror velocity of 0.05 to 0.07 cm sec^{-1} being chosen (Vidrine, 1980). It is possible to obtain spectra at higher velocities with some loss of sensitivity, and a narrow region around the Helmholtz resonance frequency can be used to obtain reasonably high signal-to-noise ratios in the corresponding spectral region. Because of the large signal phase changes around the Helmholtz frequency, care must be taken in phase-correcting this spectral region. The Helmholtz resonance frequency may be raised somewhat by reducing the cell volume, but the existing cell designs already have rather small volumes, and reducing volume further would compromise the convenience of the method with many types of samples.

A potentially important reason for obtaining pa spectra at higher modulation frequencies (mirror velocities) is depth profiling. Figure 4 shows PA-FT-IR spectra of a catalyst surface obtained at a variety of mirror velocities. Although most of the major features of the spectra are identical,

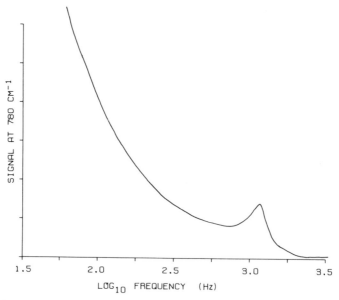

Fig. 3. Frequency response spectrum of a Gilford pa cell. Other commercial solid-sample pa cells have quite similar response curves because of the untuned acoustic design.

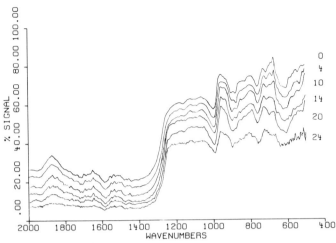

Fig. 4. PA-FT-IR spectra of a catalyst surface obtained at various mirror velocities.

small differences in relative band heights are evident. According to the theory of PAS (Rosencwaig, 1978; Betteridge *et al.*, 1979), these differences are caused by the dependence of effective penetration depth on modulation frequency. Determination of the compositional depth profile by simply obtaining pa spectra at a variety of mirror velocities could become a powerful analytical method, but no detailed studies using this method have yet been done. Some limitations exist: at the higher velocities, sensitivity is poor and longer measurement times must be used. Also, signals due to absorbing gases in the cell must be subtracted separately for each velocity to avoid artifacts. Special-purpose microvolume cells developed to maximize high-frequency response would facilitate this application. Depth profile information can also be gleaned from the phase pa spectrum (described later), especially in combination with variable modulation frequency information.

C. Effect of Sample Surface Morphology

A unique aspect of photoacoustic spectroscopy is its relative insensitivity to sample surface morphology. Transmission, reflectance, and emission spectroscopy each have strict requirements, and few solid samples can be measured without considerable sample preparation. Transmission samples must be uniform in thickness, optically thin, and free of scattering. Specular reflectance samples must either be optically thin and uniform on a reflecting substrate or very high polished (for ellipsometry). Diffuse reflectance samples must be highly scattering. Attenuated total reflectance (ATR) samples must be flat and easily deformable or very finely powdered. Emission samples must either be optically thin and uniform (with a reflecting or transmitting substrate, if any) or highly scattering. All of these samples can be measured by photoacoustic detection, but the technique is especially valuable because samples that are *not* flat, deformable, highly scattering, or optically thin can also be measured.

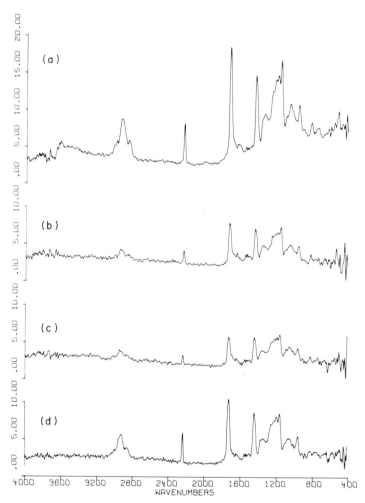

Fig. 5.   PA-FT-IR spectra of a nitrile-containing resin in different surface morphologies: (a) powder; (b) sawn surface; (c) smooth surface; (d) pellets.

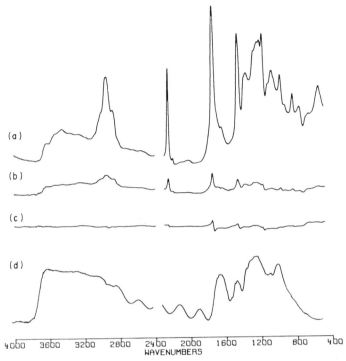

Fig. 6. Diffuse reflectance spectra of the same samples of a nitrile-containing resin as measured in Fig. 5: (a) powder; (b) sawn surface; (c) smooth surface; (d) pellets.

Figure 5 illustrates this insensitivity to sample morphology. Shown are pa spectra of several samples of a nitrile-containing resin. Their chemical compositions are identical, but their surface morphologies vary greatly. As a comparison, these samples were also measured by diffuse reflectance (Fig. 6). Note the wide variations in spectral features. The polished surface shows a pure Christiansen effect spectrum, and the powder spectrum resembles the pa spectrum. The pellet spectrum was reproduced poorly from pellet to pellet and shows puzzling features. Possibly the scattering centers have a somewhat different composition from the bulk resin? Traditionally, the rough resin sample would have to be ground at liquid N_2 temperature (cryogrinding) and mixed with powdered KBr to allow a useful spectrum to be obtained (by KBr pellet or diffuse reflectance) for determination of composition. An alternative now exists: photoacoustic measurement of the resin sample in its rough, unprepared state.

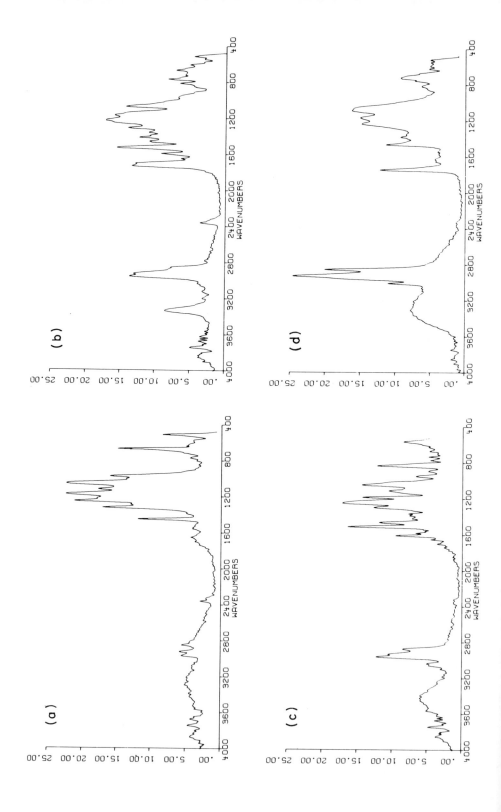

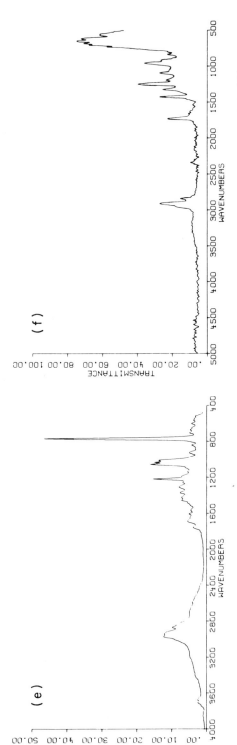

Fig. 7. PA-FT-IR spectra of various unprepared samples: (a) blue plastic; (b) polyurethane chips; (c) phenoxy pellets; (d) liquid lecithin; (e) Empirin tablets; (f) opaque panel (PVC and Filler).

V. PRESENT APPLICATIONS

A. A No-Preparation Method

Preparation of a solid sample for spectrometric analysis can be the most difficult part of the analytical procedure. Most raw samples are not polished flat, optically thin, or highly scattering; many are not meltable, dissolvable, easily powderable or polishable, or readily deformable. Photoacoustic measurement of a sample in its raw state can often be done when sample preparation for other ir spectrometric methods is difficult, uncertain, or time-consuming. Significant advances in sensitivity are still being made, but photoacoustic detection is intrinsically somewhat insen-

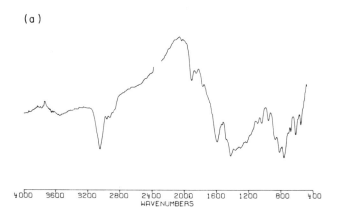

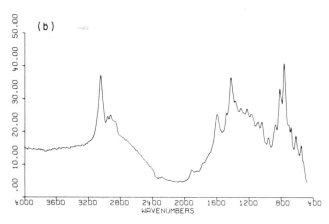

Fig. 8. (a) Diffuse reflectance and (b) PA spectra of a finely ground coal sample.

(a)

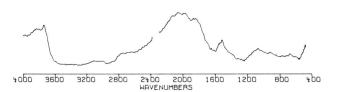

(b)

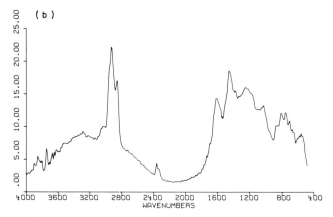

Fig. 9. (a) Diffuse reflectance and (b) pa spectra of a coarsely ground coal sample.

sitive, and measurement of high-quality spectra requires minutes, not seconds. This technique should be seriously considered for routine use whenever sample preparation, because of time requirements or possibility for error, is a more serious constraint than the measurement time required. Insoluble polymers and polymer composites represent one class of intractable samples. The materials represented by the spectra of Fig. 7 were measured without preparation. Traditional spectrometric methods would require cryogrinding for similar results. Coal is another example of a material traditionally requiring preparation. Powdered coal can be measured by diffuse reflectance, although the spectral detail obtained with pa detection is often superior (Fig. 8). Coarsely ground coal (Fig. 9) is easily measured by FT-IR/PAS, whereas other methods, including diffuse reflectance, fail (Vidrine, 1980). Another example is Rockley and Devlin's (1980) investigation of the surface aging of freshly cleaved coal by FT-IR/PAS (Fig. 10). Here, any sample preparation at all might compromise the validity of the results, and there is really no substitute for pa detection. Lowry et al. (1981) have recently measured a series of pesticide samples of FT-IR/PAS. Because the pesticide is "spraypainted" onto the surface of the clay substrate, grinding or crushing the sample would ex-

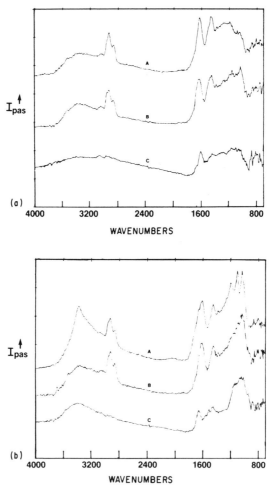

Fig. 10. PA-FT-IR spectra: (a) freshly cleaved coal surface; (b) coal surface after exposure to air.

pose large quantities of interfering clay, and the samples are not highly scattering enough for reliable diffuse reflectance analysis. However, analysis of the unprepared sample by FT-IR/PAS is straightforward, as shown in Fig. 11.

B. Investigation of Sample Preparation Artifacts

Even when sample preparation is not a difficult endeavor, the reliability of analytical results may be compromised by unknown artifacts produced by the preparation. In many cases, these artifacts or their effect on the determination can be eliminated by careful analytical design, but how does the architect of an analysis recognize the artifact without an artifact-free spectrum for comparison? With difficulty or intuition, if at all. The use of FT-IR/PAS allows spectra free of artifacts of preparation to be easily obtained for this comparison. Thus, FT-IR/PAS used in the development of an analytical procedure can improve its reliability even when pas is not used in the final routine procedure. Traditional, fairly uncomplicated, sample preparation methods, like grinding with KBr (used for KBr pellets and some diffuse reflectance work), are extremely prone to artifacts due to solid-phase ion exchange, Christiansen effects from scattering (or the lack thereof), grinding-induced pyrolysis, or simple crystalline polymorphism. With FT-IR/PAS, the options of avoiding the artifact by using pa techniques routinely for analysis or eliminating its effects by careful, sapient, analytical design are available.

C. Avoidance of Reflectance Effects

Infrared band shapes and positions are significantly affected by the Christiansen effect. Figure 12 shows the band asymmetry (especially apparent in the 1700-cm^{-1} and 830-cm^{-1} bands) in an ATR spectrum of an insulating foam. The corresponding pa spectrum shows no asymmetry (and considerably more detail in the O—H and C—H stretching bands). In diffuse and other external reflectance, these effects are often even more severe, as has been shown by the comparison in Figs. 5 and 6. Band-shape analysis in ir spectra of inorganic solids has been hampered by this type of uncertainty, and Royce's group (Laufer *et al.*, 1980) have now shown that true shapes can be measured by pa techniques.

D. Addendum

It is desirable at this time to mention further work in FT-IR/PAS that has occurred since the original writing of this chapter. Quantitative anal-

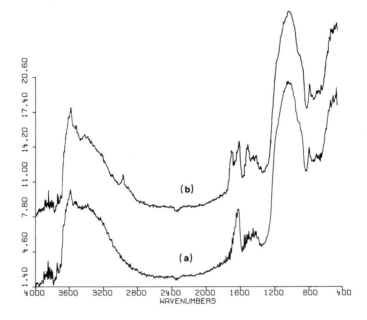

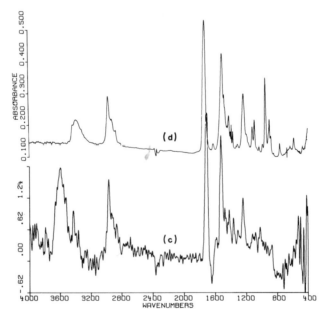

Fig. 11. PA-FT-IR spectra: (a) clay; (b) insecticide on clay substrate; (c) insecticide after spectral subtraction of the clay. For reference, (d) is a transmission spectrum of the pure insecticide.

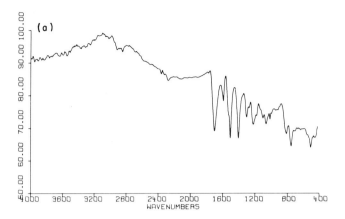

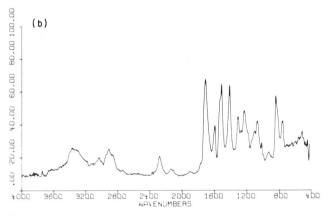

Fig. 12. (a) ATR and (b) PA-FT-IR spectra of an insulating foam sample.

ysis using FT-IR/PAS is being investigated in more detail (Rockley *et al.*, 1981; Teng and Royce, 1981; Lowry *et al.* 1981), and the trade-offs implicit in FT-IR/PAS are becoming clearer (Vidrine, 1981; Krishnan *et al.*, 1981; Yeboah *et al.*, 1981; Wisnosky, 1981; Groves *et al.*, 1981). Mehicic *et al.* (1981) have recently furnished a gallery of applications of FT-IR/PAS to petrochemical raw materials and high polymers. The FT-IR/PAS spectra of several enzymes have been measured, including horseradish peroxidase (Rockley *et al.* 1980), bovine superoxide dismutase, and laccase (Woodruff, 1981). Also, spectral differences between stellacyanin and its copper-free analog have been noted in the mid-infrared (Woodruff and Vidrine, 1981).

VI. PROSPECTIVE

A. Further Cell Development

Thus far, air at atmospheric pressure has been used for FT-IR/PAS. As pointed out in Section IV. A, cell sensitivity can be improved by using other gases, and probably by using other pressures. In current commercial cell designs, volume has been reduced as much as possible without compromising the space available for the sample unduly. However, if a flat or small sample is to be measured routinely, greater sensitivity could be achieved by reducing the cell volume further. Improving microphone performance should increase cell sensitivity, as long as external noise sources do not become predominant. A thermal mid-ir detector operated at room temperature has a theoretical maximum sensitivity of $D^* = 10^9$ cm Hz W^{-1} because of thermal noise. Pyroelectric detectors such as DTGS approach this, with $D^* = 3 \times 10^8$, but current pa cells have D^* values in the range of 10^7. Gas-microphone pa cells will never approach the theoretical maximum because of intrinsic inefficiencies in heat transfer from the sample and acoustic transfer to the microphone diaphragm. Considerable room for improvement may exist, nonetheless, and experiments with Brownian noise-limited transducers such as the different laser-microphone designs might be productive.

An alternative way to increase sensitivity and frequency response is to use a liquid instead of a gas as the acoustic transfer medium. Serious problems exist, including the high heat capacity and low thermal expansion coefficient of liquids and the imperfect transparency of liquids in the mid-ir. However, the hydrophone cell is being seriously considered as a practical future method of pa detection (Rosencwaig, 1980). A further step in this direction is the direct thermal measurement of the energy absorbed by a coating on a thermocouple or other thermal detector surface (Brilmyer et al., 1977). This is no longer photoacoustic spectroscopy, of course, but it has some possible advantages when a catalyst surface (for instance) can be directly plated onto a tiny thermoelectric transducer.

B. Future Applications

Considering some of the prognostications that have been made about pa spectroscopy since its discovery, it may be foolhardy to predict applications other than those already described. Nonetheless, some possible future applications come to mind.

Photoacoustic spectroscopy of gases was practically the only application of the pa effect before the advent of lasers, and pa photometry of gases using lasers is now an established monitoring technique. Photoacoustic detection of gases by FT-IR spectroscopy is comparable in sensitivity to transmission FT-IR with a room-temperature thermal detector in most cases, but there are advantages to pa detection in some situations. The pa spectrum is additive (peaks on a zero base line) instead of subtractive, as in a transmission spectrum. For instance, a 1% CO_2 absorption band in a gas cell may be impossible to detect in a spectrometer where the CO_2 in the purge gas may be absorbing 50% (more or less) of the light. This analysis could be simple and accurate with pa detection. A second application is in far-ir spectroscopy (Fig. 13), where Hg-arc source flicker and beam splitter vibration can be major sources of noise. Because the multiplex and throughput advantages for transmission spectra are reduced by this type of noise source, the detection of an emissionlike line spectrum could improve the spectral signal-to-noise ratio.

Two types of equilibrium processes at solid surfaces can be measured via FT-IR/PAS. First, pa detection allows the concentration of adsorbed species to be probed directly. Second, many adsorption–desorption equilibria are quite temperature-sensitive. This can enhance the photoacoustic signal from adsorbed species (Wong, 1978; Gray and Bard, 1978), and specific information about rates can be obtained from phase spectra or variation of modulation frequency.

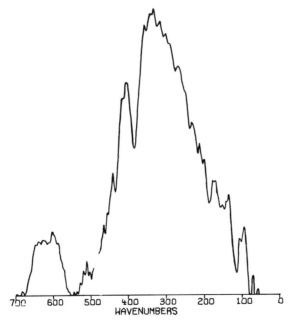

Fig. 13.   Single-beam, far-ir, photoacoustic spectrum of black pigment. A Mylar beamsplitter and polyethylene cell window are used. A resolution of 4 cm⁻¹ was used with a time of 2846.56 sec.

Robin's group (Kaya *et al.*, 1974, 1975; Robin, 1977) have extensively used modulation frequency changes and phase spectra to investigate phosphorescence and fluorescence in organic vapors. Because phase spectra are also obtainable in interferometric spectrometry, the sensitivity and high modulation frequencies available could allow experiments on species having lower quantum efficiencies or shorter lifetimes. Other relaxation processes should also be amenable to measurement using this approach.

Visible pa measurements are currently performed using mechanical choppers and monochromators. Although reasonable sensitivity is achieved by this technique, the high absorptivities characteristic of many uv–visible chromophores currently make sample preparation (dilution) essential for many solid samples (Fuchsman and Silversmith, 1979). Otherwise, pa signal saturation often occurs at the low modulation frequencies available (McClelland and Kniseley, 1976). In the uv–visible region, interferometric spectroscopy has the high modulation frequency capability required to reduce the likelihood of signal saturation and the high sensitivity necessary to offset any compromise made in pa cell sensitivity to allow detection of such high audio frequencies. The necessary FT-IR instrumentation capable of operating in the visible region exists, and such an approach only awaits the development of pa cells capable of responding to high modulation frequencies. It is worth noting that although most uv–visible detectors do not allow the realization of the normal multiplex and throughput advantages of FT-IR, photoacoustic detection is an exception: these advantages should be fully operative in uv–visible interferometric pa spectroscopy.

Although depth profiling of the solids by PAS has been investigated, both theoretically and experimentally in the uv–visible region (Adams and Kirkbright, 1976, 1977; Rosencwaig, 1978; Betteridge *et al.*, 1979; Adams *et al.*, 1979), only the most preliminary results (see Fig. 4) have been obtained in the mid-ir. Nonetheless, depth profiling of native surfaces in the low micron range is an important capability, and no other effective technique now exists for doing this on many rough-surfaced samples. Varying modulation frequency and analyzing the phase spectrum are effective methods of probing the depth profile of a solid sample, and in combination, these two techniques should allow detailed information about the sample surface and subsurfaces to be obtained.

VII. SUMMARY

Fourier transform infrared photoacoustic spectroscopy is a new, routinely usable, mid-ir measurement technique that requires minimal sample preparation and is applicable to all types of solid and liquid samples, including those difficult to measure by other techniques. The spectra are quite similar in appearance and interpretation to normal transmission spectra, so this new technique can be immediately and effectively exploited by practicing analysts who are familiar with traditional ir practice. The advantages of PAS, which have already been demonstrated in the uv–visible region, are applicable in the mid-ir, and a serious disadvantage in the uv–visible region (signal saturation) is not a serious problem in the ir region. The large multiplex and throughput advantages of modern FT-IR instrumentation effectively counteract the intrinsic insensitivity of the gas-microphone pa detector and allow PA-FT-IR spectra with good signal-to-noise ratios to be obtained in reasonable measurement times. The large range of modulation frequencies obtainable with multivelocity FT-IR instruments also makes possible specialized techniques such as depth profiling and the investigation of radiative relaxation processes. Thus FT-IR/PAS is a new ir technique that should be considered whenever sample preparation represents a difficult part of the analytical procedure, and one that cannot profitably be ignored whenever sample preparation artifacts compromise the reliability of the results. The technique also has other important capabilities, such as depth profiling, which could become major uses of PAS in the future.

REFERENCES

Adams, M. J., and Kirkbright, G. F. (1976). *Spectrosc. Lett.* **9**, 255.

Adams, M. J., and Kirkbright, G. F. (1977). *Analyst (London)* **102**, 678.

Adams, M. J., and Kirkbright, G. F., and Menon, K. R. (1979). *Anal. Chem.* **51**, 508.

Bell, A. G. (1880). *Am. J. Sci.* **20**, 305.

Bell, A. G. (1881). *Philos. Mag.* **11**, 510.

Betteridge, D., Lilley, T., and Meyler, P. J. (1979). *Fresenius Z. Anal. Chem.* **296**, 28.

Blank, R. E., and Wakefield, T. (1979). *Anal. Chem.* **51**, 50.

Brilmyer, G. H., Fujishima, A., Santbanam, K. S. V., and Bard, A. J. (1977). *Anal. Chem.* **49**, 2057.

Busse, G., and Bullemer, B. (1978). *Infrared Phys.* **18**, 255, 631.

Christiansen, C. (1884). *Ann. Phys. Chem.* **23**, 298.

Farrow, M. M., Burnham, R. K., and Eyring, E. M. (1978). *Appl. Phys. Lett.* **33**, 735.

Fuchsman, W. H., and Silverman, A. J. (1979). *Anal. Chem.* **51**, 589.

Gray, R. C., and Bard, A. J. (1978). *Anal. Chem.* **50**, 1262.

Groves, G. K., Brasch, J. W., and Jakobsen, R. J. (1981). *Proc. 1981 Int. Conf. FT-IR Spectrosc., Columbia, South Carolina.* Paper #TB11.

Kaya, K., Harshbarger, W. R., and Robin, M. B. (1974). *J. Chem. Phys.* **60**, 4231.

Kaya, K., Chatelain, C. L., Robin, M. B., and Kuebler, N. A. (1975). *J. Am. Chem. Soc.* **97**, 2153.

Kreuzer, L. B. (1977). *In* "Optoacoustic Spectroscopy and Detection" (Y.-H. Pao, ed.), Chapter 1. Academic Press, New York.

Krishnan, K., Hill, S. L., Witek, H., and Knecht, J. (1981). *Proc. 1981 Int. Conf. FT-IR Spectrosc., Columbia, South Carolina.* Paper #TB1.

Laufer, G., Huneke, J. T., Royce, B. S. H., and Teng, Y. C. (1980). *Appl. Phys. Lett.* **37,** 517.

Lloyd, L. B., Burnham, R. K., Chandler, W. L., Eyring, E. M., and Farrow, M. M. (1980). *Anal. Chem.* **52,** 1595.

Low, M. J. D., and Parodi, G. A. (1980). *Appl. Spectrosc.* **34,** 76.

Lowry, S. R., Mead, D. G., and Vidrine, D. W. (1981). *Proc. 1981 Int. Conf. FT-IR Spectrosc., Columbia, South Carolina.* Paper #TB3.

McClelland, J. F., and Kniseley, R. N. (1976). *Appl. Phys. Lett.* **28,** 467.

McDonald, F. A., and Wetsel, G. C. (1978). *J. Appl. Phys.* **49,** 2313.

Mehicic, M., Kollar, R., and Grasselli, J. G. (1981). *Proc. 1981 Int. Conf. FT-IR Spectrosc., Columbia, South Carolina.* Paper #TB2.

Michelson, A. A. (1881). *Philos. Mag. V* **31,** 256.

Robin, M. B. (1977). *In* "Optoacoustic Spectroscopy and Detection" (Y.-H. Pao, ed.), Chapter 7. Academic Press, New York.

Rockley, M. G. (1979). *Chem. Phys. Lett.* **68,** 455.

Rockley, M. G., and Devlin, J. P. (1980). *Appl. Spectrosc.* **34,** 405, 407.

Rockley, M. G., Davis, D. M., and Richardson, H. H. (1980). *Science* **210,** 918.

Rockley, M. G., Davis, D. M., and Richardson, H. H. (1981). *Appl. Spectrosc.* **35,** 185.

Rosencwaig, A. (1973a). *Opt. Commun.* **7,** 305.

Rosencwaig, A. (1973b). *Science* **181,** 657.

Rosencwaig, A. (1977). *In* "Optoacoustic Spectroscopy and Detection" (Y.-H. Pao, ed.), Chapter 8. Academic Press, New York.

Rosencwaig, A. (1978). *J. Appl. Phys.* **49,** 2905.

Rosencwaig, A., and Hindley, T. W. (1981). *Appl. Opt.* **20,** 606.

Rosencwaig, A., and Gersho, A. (1976). *J. Appl. Phys.* **47,** 64.

Royce, B. S. H., Enns, J., and Teng, Y. C. (1980). *Bull. Am. Phys. Soc.* **25,** 408.

Teng, Y. C., and Royce, B. S. H. (1981). *J. Opt. Soc. Am.,* in press.

Veingerov, M. L. (1938). *Dokl. Akad. Nauk. SSSR* **19,** 687.

Vidrine, D. W. (1979). *IR Spectral Lines* **1**(5), 2; Mead, D. G., Lowry, S. R., Vidrine, D. W., and Mattson, D. R. (1979). *Proc. Int. Conf. Infrared Millimeter Waves, 4th* (S. Perkowitz, ed.). Paper #F-4-9, IEEE Catalog #79 CH1384-7MTT.

Vidrine, D. W. (1980). *Appl. Spectrosc.* **34,** 314.

Vidrine, D. W. (1981). *Proc. 1981 Int. Conf. FT-IR Spectrosc., Columbia, South Carolina.* Paper #T4.

Wake, D. R., and Amer. N. M. (1979). *Appl. Phys. Lett.* **34,** 379.

Wisnosky, M. G. (1981). In preparation.

Wong, K. Y. (1978). *J. Appl. Phys.* **49,** 3033.

Woodruff, W. H. (1981). In preparation.

Woodruff, W. H., and Vidrine, D. W. (1981). Personal communication.

Yeboah, A., Griffiths, P., Krishnan, K., and Kuehl, D. (1981). *Proc. 1981 Int. Conf. FT-IR Spectrosc., Columbia, South Carolina.* Paper #TB4.

5

TECHNIQUES USED IN FOURIER TRANSFORM INFRARED SPECTROSCOPY

K. Krishnan

Digilab Division
Bio Rad Laboratories
Cambridge, Massachusetts

John R. Ferraro[*]

Department of Chemistry
Loyola University
Chicago, Illinois

I. INTRODUCTION

With the advent of commercial interferometers, a number of instrumental and computer techniques have been adapted to incorporate their use. Some of the instrumental techniques have also been used with dispersive instruments. It is not our goal in this chapter to compare the merits or demerits of the use of these techniques with dispersive or interferometric instrumentation. For some of these techniques, chapters have appeared in Volumes 1 and 2 of this treatise, "Fourier Transform Infrared Spectroscopy." In this chapter, we shall collect and review these techniques and shall attempt to update them with the most recent research awareness.

[*] The support of the Searle Foundation in this work is acknowledged.

FOURIER TRANSFORM
INFRARED SPECTROSCOPY, VOL. 3
Copyright © 1982 by Academic Press, Inc.
All rights of reproduction in any form reserved.
ISBN 0-12-254103-0

II. TECHNIQUES USED IN FOURIER TRANSFORM INFRARED SPECTROSCOPY

A. Instrumental Techniques

1. Attenuated Total Reflectance

Attenuated total reflectance (ATR) is a powerful technique for studying the spectra of surfaces. The samples to be studied can be free-standing films, coatings on metals, or even liquids. The ATR technique and available accessories have been thoroughly reviewed in the literature (Harrick, 1967). It is now well understood that if the absorbance-subtraction technique is to be used to compare different spectra, the maximum peak absorbances should not exceed 0.8 absorbance units. Transmission spectra of free-standing films or, in a few cases, even capillary films of liquids may yield too strong absorbances. In these cases, the ATR technique can be used as an alternative method.

The ATR accessory utilizes a prism made from a high-refractive-index material. When the infrared beam is incident on the crystal at an angle larger than the critical angle, internal reflection takes place. When a sample is placed in intimate contact with the crystal face where the reflection takes place, a standing wave is established at the crystal–sample interface and there is some penetration of the infrared radiation into the sample. The reflected radiation then has decreased intensity at those frequencies where the sample absorbs. The depth to which the radiation penetrates into the sample depends on the refractive indices of the crystal and the sample, the effective angle of incidence of the radiation, and the wavelength. The depth of penetration also depends on the uniformity of the contact between the sample and the prism.

The commonly used materials for the prism are KRS-5 and germanium. The prism has faces cut at either 45° or 60°. Table 1 (Smith, 1980) shows the depths of penetration of the infrared beam into polyethylene for different frequencies, prism materials, and prism angles. One can see that this value can range from ~0.25 μm to ~4 μm, depending on the choice of the frequencies and materials. Furthermore, the depth of penetration is lower at higher frequencies. Thus, although an ATR spectrum will resemble the transmission spectrum of the sample, the relative intensities will be different in the two spectra.

In practice, multiple instead of single internal reflections are utilized. The prism usually has a rhombohedral form, allowing samples to be placed on either side of the crystal. The crystal size normally used with dispersive instruments has the dimensions 50 × 20 × 2 mm to match the

TABLE 1[a]

Depths of Penetration for Polyethylene

Crystal	Angle (deg)	Wavenumber (cm^{-1})			
		2000	1700	1000	500
KRS-5	45	1.00	1.18	2.00	4.00
	60	0.55	0.65	1.11	2.22
Ge	30	0.60	0.71	1.20	2.40
	45	0.33	0.39	0.66	1.32
	60	0.25	0.30	0.51	1.02

[a] From Smith, 1980.

rectangular slit image of these instruments. While such crystals can be used with FT-IR instruments, large energy losses will be incurred. Instead, it is best to use crystal dimensions matched to the circular beams produced by the FT-IR instruments. The typical dimensions of the crystal used with FT-IR instruments are 50 × 3 × 2 mm. By means of suitable optics, the whole of the infrared radiation can be directed into the crystal. Figure 1 shows a typical ATR accessory. With such crystals, depending on the effective angle, between twenty and thirty reflections can take

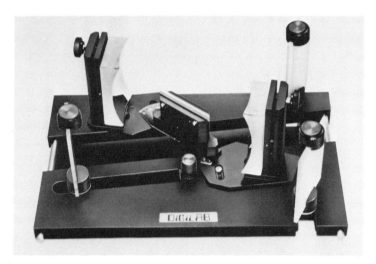

Fig. 1. Photograph of a commercial beam-condensing ATR accessory, showing a KRS-5 crystal placed between two parobolic mirrors. The dimensions of the crystal are 50 × 3 × 2 mm.

place. Because of the significant energy losses in the ATR experiment, the
throughput advantage of the FT-IR instrument becomes a significant factor. The use of a mercury–cadmium telluride (MCT) detector will further
enhance the effectiveness of the technique.

A typical ATR experiment is performed as follows: The background
reference spectrum is first obtained with the crystal alone. The sample to
be examined is then placed in intimate contact with the crystal. This can
be done by means of a suitable pressure plate arrangement. Alternatively,
the sample film can be directly deposited on the crystal (Jakobsen, 1979).
If the first method is used and if the ATR spectra of a series of samples are
to be compared quantitatively, it will be necessary to assure the same extent of contact between the ATR crystal and the sample. This can be done
in most cases by using a torque wrench and applying the same pressure in
the sampling method. The spectrum of the sample is then recorded, and it
can be ratioed against the reference spectrum to produce the ATR spectrum of the sample. Figure 2 shows the ATR spectrum of a rubber gasket
recorded for 30 sec. The ATR technique can also be used as a microsampling technique, as in Fig. 3, where the ATR spectrum of a paint chip from
an automobile is shown. In addition to paint, the ATR technique is very
useful for the study of coating, rubbers, and so on.

The ATR technique can also be used with polarized incident radiation
to study molecular ordering. Jakobsen (1979) has used this technique to
measure the molecular ordering of stearic acid adsorbed on iron. Fringeli
(1977) has used the polarized ATR technique for studying the infrared
spectra of phospholipids using dispersive instruments.

2. Diffuse Reflectance

Another sampling technique that has met with wide acceptance is diffuse reflectance. This technique can be used with powders and turbid liquids and, to a limited extent, with coatings on flat surfaces. The technique
of diffuse reflectance has been reviewed in a number of books (Hardy,
1936; Judd and Wyszecki, 1963; Billmeyer and Daltzman, 1966; Kortum,
1969; Frei and McNeill, 1973). Most of the published reports using this
technique have been confined to the uv–visible region of the electromagnetic spectrum. In this spectral region, high-intensity sources and very
sensitive detectors exist, allowing the detection of the weak, diffusely reflected radiation from the sample.

The common experimental device used in the diffuse reflectance measurements is an integrating sphere, the inside of which can be coated with
a nonabsorbing material such as magnesium oxide. The sample and the
detector can be placed on the surface of the sphere. Under these condi-

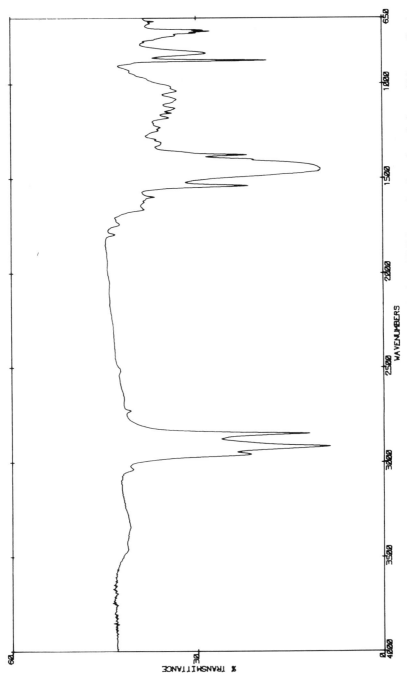

Fig. 2. ATR spectrum of a rubber gasket from a washing machine recorded using a KRS-5 crystal at 4 cm^{-1} resolution and 30-sec collection time.

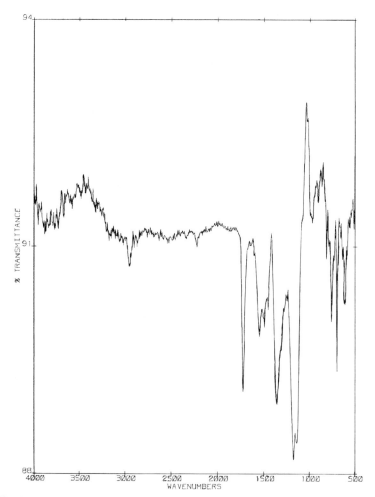

Fig. 3. ATR spectrum of a 0.5×0.5 mm paint chip from a 1975 Plymouth automobile. The spectrum was recorded at 4 cm^{-1} resolution for 2 min, showing the microsampling capabilities of the technique. A KRS-5 crystal was used.

tions, the measured spectrum is essentially independent of the positions of the sample, the detector, and the spatial distribution of the relected light.

The diffuse reflectance technique has been used with only limited success in the mid-infrared spectral region prior to the advent of FT-IR instruments. This is mainly due to the fact that dispersive instruments have limited energy throughput. To overcome this problem, specially designed accessories have been developed. These have included the Coblentz sphere (Coblentz, 1913), the Gier–Dunkle cavity (Gier *et al.*, 1954), the hemispherical collecting mirrors (White, 1964), hemiellipsoidal (Wood *et al.*, 1976), ellipsoidal (Wood *et al.*, 1976; Körtum and Delfs, 1964; Dunn *et al.*, 1966), and integrating spheres (Dunn, 1965; Willey, 1976). Of these, the hemielliptical, or ellipsoidal, mirror arrangements are the most efficient in collecting the diffusely reflected radiation and directing it toward the detector.

Using some of these devices and dispersive instruments, the feasibility of using the diffuse reflectance technique in the mid-infrared region for chemical spectroscopy and catalyst studies has been demonstrated (Kortüm and Delfs, 1964). Willey (1976) was among the first to use an integrating sphere in conjunction with an FT-IR instrument. However, the integrating sphere is very inefficient, and very long measurement times were needed to obtain quality spectra of good signal-to-noise ratio even at moderate spectral resolution. Griffiths and his co-workers (1978; Fuller and Griffiths, 1978a) have designed a fairly efficient diffuse reflectance accessory employing large ellipsoidal mirrors and have used it in conjunction with a commercial rapid-scanning Fourier transform instrument. They have demonstrated the applicability of this technique for a wide range of samples. Krishnan *et al.* (1980) have also used a commercial diffuse reflectance accessory to record the FT-IR spectra of a wide variety of samples.

A general theory for diffuse reflectance at scattering layers has been proposed by Kubelka and Munk (1931; Kubelka, 1948) for defining the relation between diffuse reflectance and the sample concentrations. For an infinitely thick layer, the Kubelka–Munk equation may be written as

$$f(R_\infty) = (1 - R_\infty)^2/2R_\infty = k/s \qquad (1)$$

where R_∞ is the absolute reflectance of the layer, k the molar extinction coefficient, and s a scattering coefficient. When the reflectance is measured with respect to a reference standard, we can write

$$R_\infty = R_\infty(\text{sample})/R_\infty(\text{reference}) \qquad (2)$$

The standard chosen should be such that it is nonabsorbing and exhibits high diffuse reflectance throughout the region of interest. Fuller and Griffiths (1978a,b) have investigated the use of powdered KBr, gold, KCl, and so on, as possible reflectance standard materials. They conclude that finely divided KCl has the fewest interferences and the highest reflectance in the infrared. Powdered KBr has also been extensively used in diffuse reflectance studies (Krishnan *et al.*, 1980).

The reflectance of the powdered alkali halides is dependent on the particle size. As the particles become larger, reflectance at higher frequencies falls off. The Kubella–Munk theory predicts a linear relationship between the molar absorption coefficient and the value of $f(R_\infty)$ in Eq. (1), provided s remains constant (s depends on the particle size). It is important, when comparing the diffuse reflectance spectra of a series of similar compounds, that the particle size be kept constant.

Alkali halide powders are not only used as the standard reference materials. It has been observed that the best diffuse reflectance spectra are obtained when the sample to be studied is mixed in with the alkali halide powder. For dilute samples in nonabsorbing matrices, it has been shown (Kortum *et al.*, 1963) that

$$k = 2.303\epsilon c \qquad (3)$$

where ϵ is the molar absorptivity and c the concentration. Then

$$f(R_\infty) = (1 - R_\infty)^2/2R_\infty = c/k' \qquad (4)$$

where $k' = s/2.303\epsilon$. Hence, the diffuse reflectance spectrum of a sample studied as a dilute solution in an inert matrix such as KCl or KBr should be very similar to that of an absorbance spectrum.

Figure 4 shows a photograph of a typical diffuse reflectance accessory, which was designed for use with the Digilab FT-IR instruments. Its configuration is such that most of the specularly reflected radiation is not picked up by the collection optics. The powdered samples under study are kept in small metal cups. Sampling cups of different depths can be used. The infrared beam can penetrate 3–5 mm into the sample.

The diffuse reflectance technique can be a very powerful microsampling method, as has been shown by Fuller and Griffiths (1978a,b). It is necessary to use the purest possible alkali halide powder in diffuse reflectance studies. The best way of assuring this is to use fresh chunks of alkali halide crystals (which can be obtained commercially from Harshaw Chemicals, Solon, Ohio) and powder them when necessary. A Wig-L-Bug (Crescent Dental Manufacturing Co., Chicago, Illinois) can be used for this purpose. Particles of different sizes can be sorted out using the Sonic Sifter (ATM Corp., Milwaukee, Wisconsin).

For studying the diffuse reflectance spectra of a series of related compounds, it is best to use the same stock of powdered alkali halide. The samples to be studied can then be simply mixed with the stock KBr powder. Alternatively, the sample could be dissolved in a suitable volatile solvent and sprayed onto the stock alkali halide powder.

Since diffuse reflectance spectroscopy is an inefficient process, the fraction of incident energy that arrives at the detector is quite small. The use of a MCT detector, instead of the normal deuterated triglycine sulfate

Fig. 4. Photograph of a commercial diffuse reflectance accessory. The samples are held in the sample stage shown in the center.

(DTGS) detector, will allow the recording of a spectrum in very short measurement times.

A typical diffuse reflectance spectrum is measured using the following sequence: First, the background spectrum using the stock alkali halide powder is recorded. The spectrum of the dilute sample in the alkali halide powder is recorded next. The samples can be kept undiluted if necessary, but the best-looking spectra are obtained when the sample concentration in the alkali halide matrix is $\sim 5-10\%$. The reflectance spectrum R_∞ is then the ratio of the sample to the reference spectrum. The Kubelka–Munk function can be obtained from R_∞ by Eq (4). Figure 5 shows the diffuse reflectance spectrum (Kubelka–Munk function) of regular and decaffeinated coffee. As an example of cesium iodide powder used to obtain diffuse reflectance spectra down to 200 cm^{-1}, Fig. 6 shows the diffuse reflectance spectrum recorded using CsI optics of 2-benzothiazolethiol.

The diffuse reflectance technique is an easy means of obtaining the infrared spectra of solids. It is particularly useful for samples such as coals and minerals that are difficult to study as KBr pellets. It is difficult to

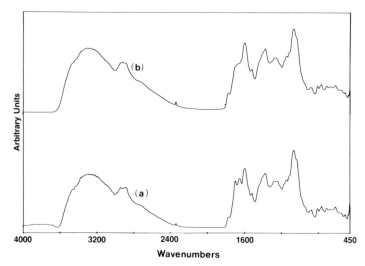

Fig. 5. Diffuse reflectance spectra of (a) regular and (b) decaffeinated freeze-dried coffees. The spectra were recorded at 4 cm¹ resolution for 4 min. One can clearly see the caffeine bands in the regular coffee spectrum. KBr was the matrix powder.

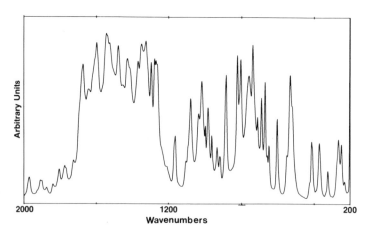

Fig. 6. Diffuse reflectance spectrum of 2-benzothiazolethiol in CsI powder recorded down to 200 cm^{-1} at 4 cm^{-1} resolution for a measurement time of 8 min. A CsI beam splitter and a room-temperature DTGS detector were employed in the measurements.

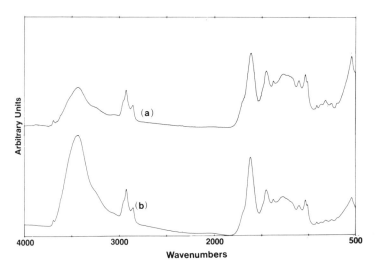

Fig. 7. A comparison between (a) the KBr pellet and (b) the diffuse reflectance spectra of the same coal sample in KBr powder. Note the cleaner base line and the sharper spectral features in the diffuse reflectance spectrum. Reprinted from *Am. Lab.* **12**(3), 104 (1980). Copyright 1980 by International Scientific Communications, Inc.

grind such samples into fine, uniform powders to produce scattering-free KBr pellets. Poorly prepared pellets yield spectra showing strong Christiansen effects. On the other hand, the diffuse reflectance spectra of such samples appear to be free of such effects. Figure 7 shows a comparison between the KBr pellet and the diffuse reflectance spectra of a coal sample. The spectra were obtained from the same sample stock. It can be seen that the diffuse reflectance spectrum has a cleaner base line and shows clearer spectral details than the KBr pellet does.

Figure 8 shows an example of absorbance subtraction using the diffuse reflectance spectra (Kubelka–Munk functions). Absorbance subtraction can be performed readily and can be a powerful qualitative analytical tool. There are still unanswered questions as to whether the diffuse reflectance technique can be used quantitatively. All the experimental evidence available to date would seem to indicate that the technique can be used quantitatively only over narrow concentration ranges, if at all.

The diffuse reflectance technique can be a very powerful technique for studying the infrared spectra of adsorbed species. Special vacuum-sealed cells can be constructed that will allow the spectra of adsorbed gases on different substrates to be recorded as well as those of air-sensitive materials. Examples of diffuse reflectance spectra of adsorbed species have been given by Fuller and Griffiths (1978a) and Krishnan *et al.*, (1980).

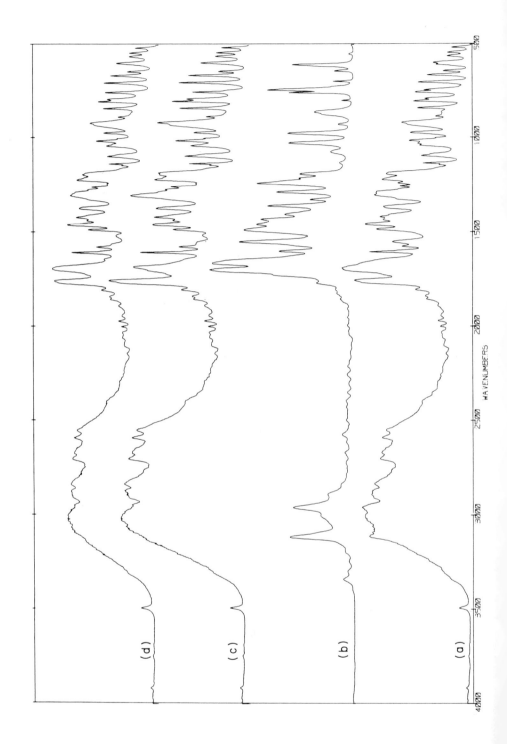

WAVENUMBERS

(a)

(b)

(c)

(d)

3. Reflection–Absorption Measurements

There are certain types of samples such as coatings on rigid metal plates that are not amenable to study by internal reflection, or ATR, techniques. Such rigid plates cannot be pressed into intimate contact with the ATR crystal and may not yield good-quality spectra. In such cases, external reflection, referred to as the reflection–absorption measurement can be used. Either single or multiple external reflection techniques can be used. The angle of incidence can be small (near normal incidence) or large (grazing angle incidence). Greenler (1966) has shown that for a single reflection the optimum angle is around 88°. He has also shown that if the multiple external reflection technique is used, there is a specific number of reflections for maximum band intensity.

For most routine applications involving external reflection measurements, a single reflection will suffice. Figure 9 shows the reflection–absorption spectrum of a polymer coating on aluminum. The angle of incidence is around 20°. Figure 10 shows the spectrum of a grinding wheel clearly exhibiting the bands due to the organic binders. As has been pointed out by Jakobsen (1979), if larger angles of incidence are required, any commercial ATR accessory can be modified for use as an external reflection attachment. Jakobsen (1979) has shown a few examples of external reflection FT-IR spectra.

When we desire to study the reflection spectrum of a very thin coating on a metal surface, or adsorbed species, the technique just described may not suffice. When light is reflected from the surface of a metal, the incident and reflected waves combine, in most cases, to form a standing wave at the surface. This standing wave will have a node at the surface and may not show significant interaction with the sample. However, when the radiation is incident at nearly grazing angles and is also polarized with the electric vector parallel to the plane of incidence, the standing wave will have a large amplitude at the surface (Greenler, 1966; Francis and Ellison, 1959). This will lead to maximum interaction between the sample and the radiation, and good-quality spectra can be obtained. The sensitivity of the technique can be enhanced by using multiple, polarized, external reflections. Under certain circumstances, this technique can also allow the layer thickness to be measured. Thus the external reflection technique

---

Fig. 8. The absorbance-subtraction technique applied to diffuse reflectance spectra. Spectra recorded at 2 cm^{-1} resolution and measurement time around 8 min for each spectrum. The spectra are presented in the Kubelka–Munk format: (a) mixture of acetylsalicylic acid and caffeine; (b) pure caffeine; (c) the difference; and (d) pure acetylasalicylic acid. Note the almost exact correspondence between the two spectra of acetylsalicylic acid (c and d).

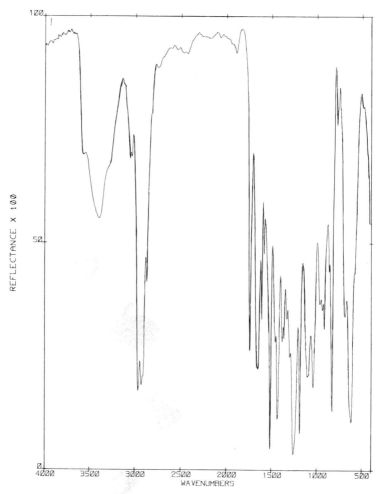

Fig. 9. Specular reflectance spectrum of polymer coating on aluminum. The reference was a front surface mirror. The spectrum was recorded at 4 cm^{-1} resolution data collected for 30 sec.

can be used to study monomolecular layers, coatings, and corrosion products on metal surfaces. Figure 11 is a photograph of a commercially available, single-reflection, polarized, grazing angle reflectance accessory.

Handke *et al.* (1980) have published the spectra of phosphates on iron obtained using an FT-IR instrument. Ishitani *et al.* (1980) have used the polarized reflection technique to study the infrared spectrum of oxide layers on metallic copper. Figure 12 shows the polarized reflection spectrum of a very thin layer of silicone grease on aluminum foil. Figure 13 shows the polarized reflection spectrum of a monolayer of an organic coating on brass.

4.   Microsampling Techniques

Fourier transform infrared spectroscopy with its enormous throughput advantage allows for relatively easy microsampling studies. The micro-

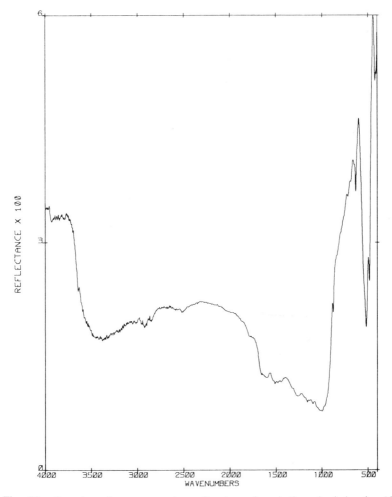

Fig. 10.   Specular reflectance spectrum of a piece of a grinding wheel showing Al_2O_3 bands and some bands due to an organic binder. The measurement conditions were similar to those in Fig. 9.

Fig. 11. A photograph of a commercial, polarized, grazing angle reflectance accessory. The unit incorporates a gold wire grid polarizer and the angle of incidence is ~75° from the normal.

sample under study can be mounted on a pinhole with suitable beam-condensing optics. Figure 14 shows a spectrum of polystyrene recorded through a 50-μm pinhole. The amount of polystyrene seen by the infrared beam amounts to around 90 ng. Using special microsample handling techniques (Anderson and Wilson, 1975), Cournoyer *et al.* (1977) have published the spectrum of 90 pg of triphenylophosphate in cellulose acetate using an FT-IR instrument. Lacy (1979) has been able to record the FT-IR spectra of contaminants in the subnanogram levels. The microsampling technique can be used to obtain the spectra of environmental particulates, tiny defects on coatings, and so on. The use of the diamond anvil cell, discussed later, represents another type of microsampling technique.

5. Emission Spectroscopy

The technique of infrared emission spectroscopy and the principles behind it have been discussed in detail by Bates (1978) and will not be repeated here. Emission spectroscopy is particularly useful for the study of samples that would be hard to do otherwise. These may be samples at very high temperature (Bates and Boyd, 1973); remote samples, such as emissions from smokestacks (Low and Clancy, 1967); or very thin films, such as monolayers or corrosion products on metals.

As with any other FT-IR sampling technique, one needs a reference spectrum; in the case of emission it can simply be that of a blackbody. The sample under study is usually a graybody, and its spectral emissivity

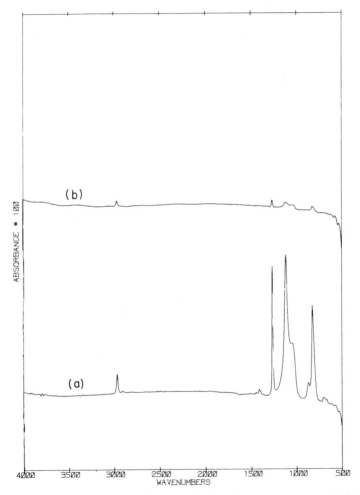

Fig. 12. Reflectance spectrum of a very thin film of silicone grease on aluminum foil. The spectra are plotted as the logarithm of reflectance using a front surface aluminized mirror as the reference. The spectra are: (a) incident electric vector parallel to the plane of incidence and (b) electric vector perpendicular to the plane of incidence. The spectra are plotted on the same scale, the full height of the plot being 1.6 absorbance unit. The spectra were recorded at 4 cm⁻¹ resolution for a measurement time of 4 min/spectrum.

can be defined as

$$E_\nu = J(\nu,\, T)/J_B(\nu,\, T)$$

that is, the ratio of the radiancy of the graybody and the blackbody. From Kirchhoff's law $E_\nu = \alpha_\nu$, where α_ν is the spectral absorptivity. For normal solids and liquids that are partially transparent in the infrared, theory has been worked out (Bates, 1978) to take into account the reflection of

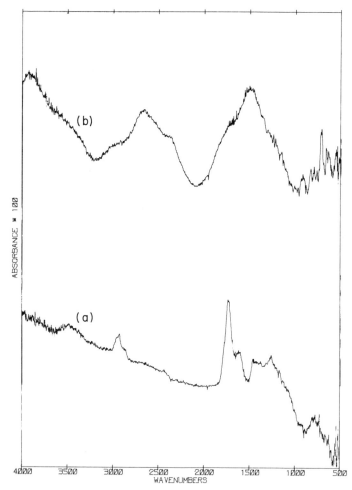

Fig. 13. Reflectance spectrum of an organic coating (monolayer on brass). The measurement conditions were the same as in Fig. 12. The full height of the plot corresponds to 0.08 absorbance unit. (b) Note the complete absence of the carbonyl bands when the incident electric vector is perpendicular to the incident plane.

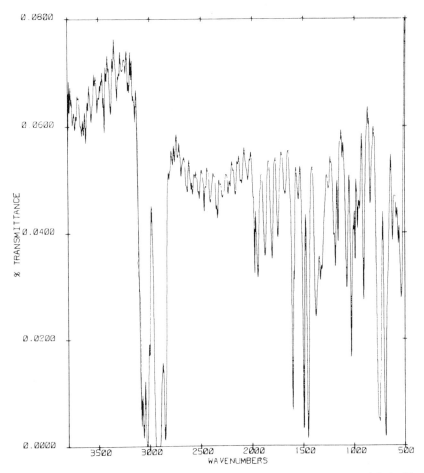

Fig. 14. Unsmoothed spectrum of polystyrene recorded through a 50-μm pinhole. The spectrum was recorded with a spectral resolution of 8 cm^{-1}, and the measurement time was around 8 min. Note that the maximum transmittance of the pinhole is less than 0.08%.

the emitted radiation within the sample, at the sample–air interface, and so on. In such a case it can be shown that

$$E + r + t \simeq 1$$

where E is the emissivity, r the reflectivity, and t the transmissivity.

In performing emission measurements, the emitting sample becomes the source for the FT-IR instrument. The radiation from the sample is collimated by means of suitable optics and directed into the interferometer. A room-temperature DTGS or a liquid-nitrogen-cooled MCT detector can be used. The reference, or background, spectrum is obtained using a blackbody held at the same temperature as the sample to be measured. Different blackbodies have been used by different investigators (Hadni, 1967; Rodgers, 1972; Barr, 1969; Kember and Sheppard, 1975; Griffiths, 1975). For most qualitative work, a good approximation to a blackbody can be obtained by using a metal plate with flat black paint on it (Kemper and Sheppard, 1975; Griffiths, 1975). The sample and the blackbody can be arranged in such a manner that one of them can easily swing into the place of the other. The ratio of the emissions of the sample and the reference blackbody gives the emission spectrum of the sample.

Emission spectra of samples, particularly solids, can be obtained even at room temperature. The emission will increase with an increase in temperature. High-quality emission spectra of most solid samples can be obtained in temperature ranges from 50 to 150°C. In doing emission studies, it is important to keep the sample thickness as small as possible. If the sample is fairly thick, temperature equilibrium will be hard to attain, i.e., the interior of the sample may be at a different temperature than the surface. Also, self-absorption processes involving emission from the bulk of the sample may take place, leading to a featureless spectra. Using samples in the form of very thin films will eliminate much of this problem. Powder samples may be ground up into a fine paste using Nujol. A thin film of this paste is then spread on an aluminum foil, yielding excellent emission spectra. Figure 15 shows an example of a spectrum of $Co(NO_2)(NH_3)_2Cl_2$ in Nujol.

In some cases, particularly when the sample is a very weak emitter, the background emission from the beam splitter and the walls of the spectrometer may become comparable to the emission from the sample. This may lead to distortions and the appearance of spurious features in the recorded spectra. A method for eliminating the background has been outlined by Kember et al. (1979). In this case one records three interferograms: one for the sample, one for the reference blackbody, and the third one for a highly reflecting (therefore poorly emitting) front surface mirror replacing the sample. Sample and reference spectra that are free of the background emission can be obtained by subtracting the third recorded interferogram from each of the first two. These two interferograms can then be Fourier transformed to give the resulting emission spectrum. For more specific details on the emission technique, the reader is encouraged to consult Bates (1978).

6. Fourier Transform Far-Infrared Spectroscopy

One of the major advantages of Fourier transform spectroscopy is the ease with which far-infrared spectra can be obtained. The field of far-in-

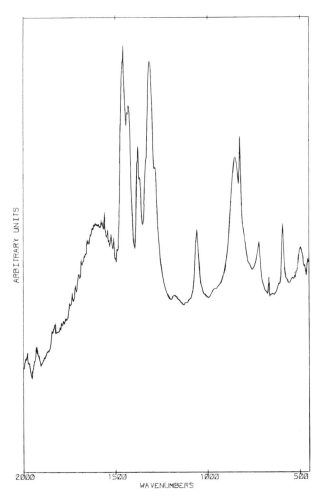

Fig. 15. The emission spectrum of the metal complex $Co(NO_2)(NH_3)_2Cl_2$. The sample was ground up with Nujol and smeared on an aluminum foil and heated to about 60°C. The data collection time was around 30 sec at 4 cm^{-1} resolution. The reference was a blackbody (black paint on aluminum) at the same temperature.

frared spectroscopy has been reviewed previously by Durig and Cox (1978). For work in the spectral region from 400 to 10 cm^{-1}, the KBr optics in the FT-IR instrument must be replaced by a Mylar beam splitter and a suitable detector. Either a room-temperature DTGS detector with high-density polyethylene windows or a liquid-helium-cooled germanium bolometer can be used. Since most alkali halides do not transmit far-infrared radiation, sampling materials such as lenses and windows have to be made of different materials. The material of choice for windows in the far-infrared region is high-density polyethylene, or polypropylene. Figure 16 shows the far-infrared spectrum of high-density polyethylene. For certain applications below 200 cm⁻¹, crystalline quartz can be used. The spectral region that can be covered in the far infrared depends on the thickness of

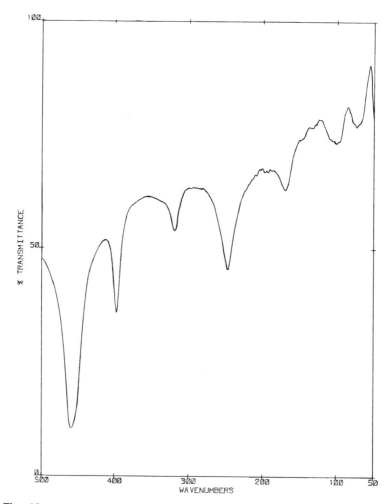

Fig. 16. Far-infrared transmission spectrum of high-density polyethylene. A globar source, 6.25-μm Mylar beam splitter, and a room-temperature DTGS detector were employed and the spectrum was recorded for a measurement time of around 8 min at 4 cm⁻¹ spectral resolution. Dry nitrogen was used for purging the FT-IR instrument.

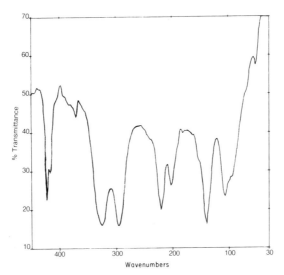

Fig. 17. The far-infrared transmission spectrum of the metal complex $Zn(pyrazine)_2Cl_2$. The sample was ground up with Nujol and held between two high-density polyethylene windows. The measurement time was 4 min, and the other experimental conditions were the same as in Fig. 16.

the Mylar beam splitter. For routine survey spectra, the region between 450 and 50 cm^{-1} can be covered using a 6.25-μm-thick Mylar film and the standard mid-ir source (water-cooled globar). For the region lower than 50 cm^{-1}, thicker beam splitter such as 25, 50, or 125 μm may be used. In the latter case, a high-pressure mercury lamp should be used as the source.

The spectra of solids in the far infrared can be recorded either as a Nujol mull or in a polyethylene pellet. Nujol is transparent in the far infrared. If the Nujol mull is held between two polyethylene plates, the use of the blank polyethylene windows as the reference will eliminate the features due to the polyethylene seen in Fig. 16. Figure 17 shows the far-infrared spectrum of a metal complex, $Zn(pyrazine)_2 Cl_2$, in a Nujol mull.

Gas-phase work in the far infrared can be performed easily using either 10-cm single-pass or multipass long-pathlength cells. The cell windows can be either high-density polyethylene or polypropylene. Durig and Cox (1978) have obtained excellent far-infrared spectra of a number of gases recorded at reasonably high resolution. Liquid samples can also be held in molded polyethylene cells. Ferraro and Basile (1979b) have shown that the diamond anvil cell can be effectively used in the far-infrared region.

7. Fourier Transform Infrared Spectroscopy
   with the Diamond Anvil Cell

The diamond anvil cell (DAC) has demonstrated its versatility during the past 20 years (Lippincott *et al.*, 1960, 1961; Weir *et al.*, 1959; Ferraro, 1971; Ferraro and Basile, 1974, 1979a,b; Adams and Payne, 1972). It has been used in combination with optical and vibrational spectroscopy, both as a microanalytical tool (Ferraro and Basile, 1979b) as well as a pressure cell (Lippincott *et al.*, 1960, 1961; Ferraro, 1971; Ferraro and Basile 1974, 1979b; Adams and Payne, 1972). The first use of the DAC for sampling purposes was made by Weir *et al.* (1959), who interfaced the cell to a dispersive spectrophotometer in the mid-infrared region. The use of the cell in the far-infrared region (< 200 cm^{-1}) was first demonstrated by Ferraro and co-workers [Ferraro *et al.* (1966) and Postmus *et al.* (1968)].

Due to the small aperture in the DAC (smallest diamonds can be ~ 0.1 mm^2 in area) and the critical angle of diamonds, it becomes necessary to condense the source beam for infrared studies. For a discussion of the problems involved in the use of diamond windows in the DAC, see Adams and Sharma (1977). For dispersive instruments, a 4-to-6× beam condenser has been used by Weir *et al.* (1959), Brasch (1965), Jakobsen *et al.* (1970), Ferraro *et al.* (1966), and Postmus *et al.* (1968). In the case of interferometric measurements, a light pipe has been used successfully (McDevitt and co-workers, 1967). Additionally, a Perkin-Elmer 4× beam condenser has been coupled with the Digilab model 14 (R. J. Jakobsen, Private communication), as well as the Bruker 1FS 114e (Klaeboe and Woldback, 1978). Recently, a 6× Harrick beam condenser was interfaced with the Digilab FTS-20A and FTS-20 interferometers (Ferraro and Basile, 1980); Krishnan *et al.*, 1980). For the measurement of emission interferometry at high pressure, the reader is referred to the review article by Lauer (1978), which deals with high-pressure infrared interferometry). This section will attempt to update the work in this area and will present newer applications of the technique.

In obtaining spectra in the DAC, the transmission spectrum of the sample was recorded by ratioing the spectrum of the diamond cell with the sample against that of the blank cell. Figure 18 shows the comparison between the single-beam spectra with and without the diamond cell in the infrared beam. Figure 19 shows an expanded view of the single-beam spectrum through the diamond cell. The strong absorptions between 2600 and 2000 cm^{-1} are plainly visible and limit the use of the cell in this region.

Krishnan and co-workers (1980) demonstrated the use of the DAC with the Digilab FTS-20 in the mid-infrared region. Very practical applications were demonstrated.

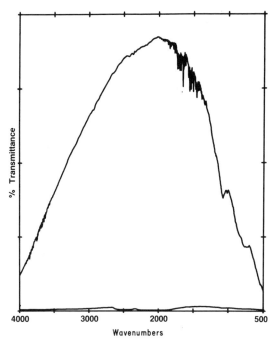

Fig. 18. Comparison of single-beam spectrum with and without the diamond anvil cell infrared beam. Reprinted from *Am. Lab.* **12**(3), 104 (1980). Copyright 1980 by International Scientific Communications, Inc.

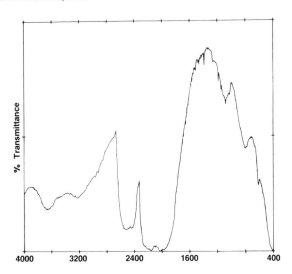

Fig. 19. Expanded view of single-beam spectrum through the diamond cell. Reprinted from *Am. Lab.* **12**(3), 104 (1980). Copyright 1980 by International Scientific Communications, Inc.

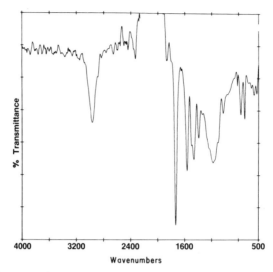

Fig. 20. Spectrum of water-based acrylic enamel paint chip in diamond cell. Reprinted from *Am. Lab.* **12**(3), 104 (1980). Copyright 1980 by International Scientific Communications, Inc.

Figures 20, 21, and 22 show the diamond cell spectra of three different paint chips. The spectra were recorded at 4 cm^{-1} resolution for a measurement time of about 4 min/sample. Figure 20 shows the spectrum of a water-based acrylic enamel, Fig. 21 that of a nonaqueous dispersion (NAD) acrylic enamel, and Fig. 22 that of an acrylic solution lacquer. The

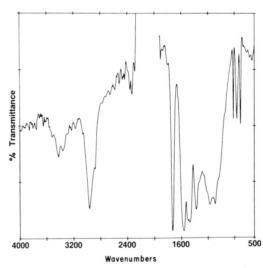

Fig. 21. Spectrum of NAD acrylic paint chip in the diamond cell. Reprinted from *Am. Lab.* **12**(3), 104 (1980). Copyright 1980 by International Scientific Communications, Inc.

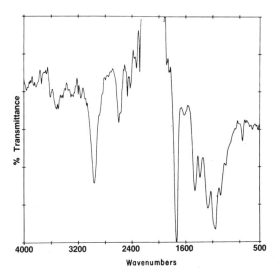

Fig. 22.   Spectrum of an acrylic solution lacquer paint chip in the diamond cell. Reprinted from *Am. Lab.* **12**(3), 104 (1980). Copyright 1980 by International Scientific Communications, Inc.

spectra exhibit very high signal-to-noise ratios and are clearly distinguishable from one another.

Figure 23 shows the transmission spectrum of a single strand of human hair recorded through the diamond cell. Finally, Fig. 24 shows the use of absorbance-subtraction techniques using the diamond cell. The spectra

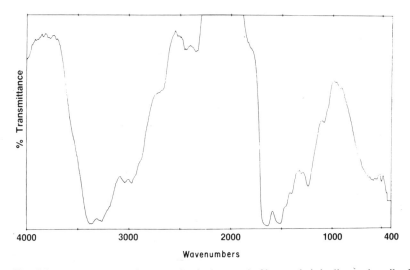

Fig. 23.   Transmission spectrum of a single strand of human hair in diamond anvil cell. Reprinted from *Am. Lab.* **12**(3), 104 (1980). Copyright 1980 by International Scientific Communications, Inc.

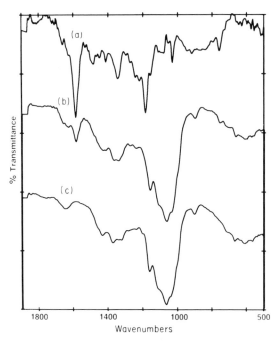

Fig. 24. Absorbance-subtraction technique using the diamond anvil cell. (a) spectrum of stock paper; (b) spectrum of stock paper with spot of ink; (c) spectrum of difference. Reprinted from *Am. Lab.* **12**(3), 104 (1980). Copyright 1980 by International Scientific Communications, Inc.

presented are those of stock paper, a different region of the paper with a spot of ink on it, and the difference spectrum. The difference spectrum shows only the features caused by the ink. These examples illustrate the use of the diamond cell as a microsampling analytical device in the mid-infrared region using an interferometer.

The application of this technique in an energy-starved region, such as the far infrared (FIR), is demonstrated in the following figures. The work is that of Ferraro *et al.* (1980; Ferraro and Basile, 1980). Using 12.5- and 25-μm Mylar beam splitters, the useful region of the FIR was extended to 25 cm^{-1}. This accessibility allows new applications to surface. For example, AgI-type ionic conductors have ν_{AgI} modes at about 100 cm^{-1}, and these can be readily studied. It may also be possible to examine soft modes in other conductors in the region less than 50 cm^{-1} as well as other very low energy modes in other molecules with pressure. It should be cited that Klaeboe and Woldback (1978) have used similar techniques and obtained spectra of *trans*-1,4-diiodocyclohexane at 41 and 57 cm^{-1}.

Figure 25 illustrates the spectrum to 4-picolinium Ag$_5$I$_6$, a new ionic conductor (Ferraro *et al.*, 1980). Figure 26 demonstrates the sensitivity to pressure of a breathing-type vibration ($\nu_{sym}SnCl$) in ($\phi_4As)SnCl_3$ (Postmus *et al.*, 1967). The spectrum of β-AgI is depicted in Fig. 27. Figure 28 shows the spectrum of yellow HgO with absorption at 60 cm^{-1}. Figure 29 illustrates the spectrum of the AgI mode at 107 cm^{-1} in pyridinium Ag_5I_6 as a function of pressure (Ferraro *et al.*, 1980).

In conclusion, the spectra presented in this section illustrate the ease with which the infrared spectra of a variety of difficult samples may be obtained throughout the infrared region using an interferometer–DAC linkup. The reader should be made aware of the fact that the present commercial dispersive infrared spectra are not amemable for use below 180 cm^{-1}.

8. Photoacoustic Spectroscopy

Another technique for studying the spectra of gases, liquids, or solids has recently become popular and is known as the photoacoustic spectros-

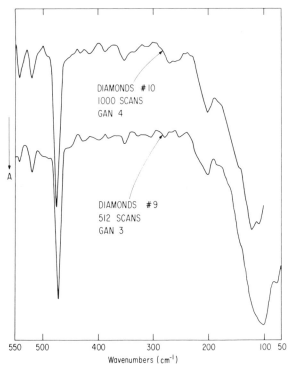

Fig. 25.   Spectra of 4-picolinium Ag_5I_6 in the diamond anvil cell. From Ferraro and Basile (1980).

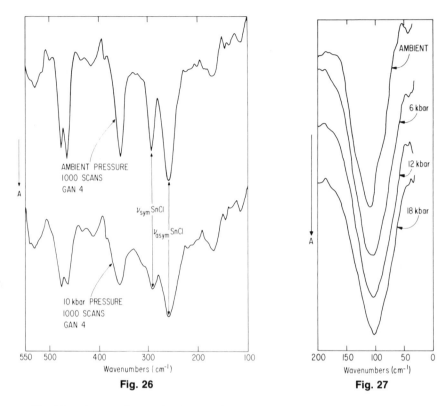

Fig. 26. Spectra of $(\phi_4 As)SnCl_3$ at ambient and 10-kbar pressure using #10 diamonds [see Postmus *et al.* (1967)]. From Ferraro and Basile (1980).

Fig. 27. Spectra of β-AgI as a function of pressure. From Ferraro *et al.* (1980).

copy technique (PAS). Instead of using the conventional pyroelectric or photovoltaic detectors, in this technique one uses the photoacoustic detector. The photoacoutic (also referred to as the optoacoustic) technique was first developed by Alexander Graham Bell and other investigators in the 1880s (Bell, 1881; Tyndall, 1881; Röntgen, 1881). A gas, for example, is placed inside a sealed cell and is illuminated with chopped or pulsed radiation. Any energy absorbed by the gas is converted wholly or partly into kinetic energy, leading to pressure fluctuations inside the cell. These pressure fluctuations can be detected by a sensitive microphone placed inside the sealed chamber. Thus the enclosed cell containing the gas sample and the microphone acts as a detector for the radiation. Figure 30 shows a typical layout of a photoacoustic detector. The detector chamber can accept samples in any state—gas, liquid, or solid. A suitable window

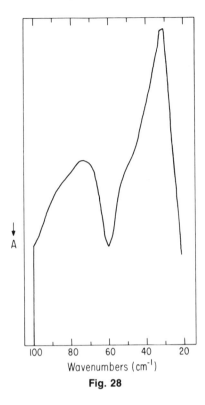

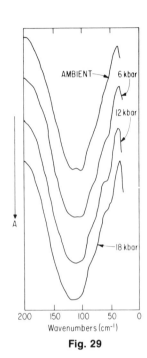

Fig. 28

Fig. 29

Fig. 28. Spectrum of yellow HgO at ambient pressure.
Fig. 29. Spectra of $(PyH)Ag_5I_6$ as a function of pressure. From Ferraro *et al.* (1980).

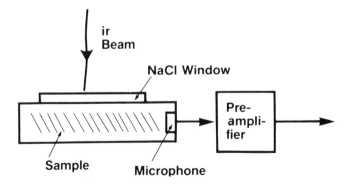

Fig. 30. Typical layout of a photoacoustic detector.

is provided for allowing the radiation to be incident on the sample. The cell contains a very sensitive microphone. Suitable gas transfer tubes allow the cell to be purged or different gases to be introduced.

Most of the earlier reported work involving the photoacoustic technique was done in the uv–visible region of the electromagnetic spectrum. Rosencwaig (1975) has reviewed some of this work and has also shown that the technique can be used effectively with solids. When working with solids, in addition to the sample, an inert gas is also introduced with the cell. The photoacoustic signal in this case arises from the periodic heat flow from the solid to the surrounding gas as the solid is heated by the pulsed radiation. This periodic heat flow produces corresponding pressure fluctuations in the cell that are detected by the microphone. The signal produced by the photoacoustic detector is proportional to the heat coming off the sample. Both the magnitude and phase of the acoustic signal are dependent on the thermal properties of the sample and the surrounding gas as well as on the modulation frequency of the incident electromagnetic radiation. The PAS spectrum corresponds qualitatively to the absorbance spectrum of the sample, assuming that dominant nonradiative processes are involved in converting the absorbed electromagnetic energy into thermal energy. Rosencweig and Gersho (1976) have developed a theory of photoacoustic spectroscopy and have shown that, under suitable experimental conditions, the PAS signal will be proportional to

$$\theta = I_0\alpha(\lambda)/4\pi f\rho C$$

where θ is the amplitude of the temperature oscillations induced in the sample surface by the incident electromagnetic radiation of modulation frequency f and intensity I_0 at wavelength λ, ρ is the density, and C the specific heat of the sample material. This theory shows that, within a wide range of experimental conditions, a photoacoustic spectrum may be identical to the conventional absorption spectrum of the sample. In this linear region the photoacoustic signal is inversely proportional to the modulation frequency. Photoacoustic saturation occurs when $\mu \gtrsim \alpha(\lambda)^{-1}$, where $\mu = (k/\pi f\rho C)^{1/2}$ is the diffusion length of the material and k its thermal conductivity. Saturation leads to loss of spectral contrast since the radiation is absorbed in surface layers deeper than the diffusion length.

The photoacoustic technique has been extended to the infrared region of the electromagnetic spectrum only during the past couple of years. Kanstad and Nordal (1977, 1979) have used a tunable, low-power CO_2 laser to study the infrared spectra of a number of solids and have developed special PAS cells for the study of surfaces. Low and Parodi (1980a,b) have used the PAS technique in conjunction with dispersive infrared spectrometers. With dispersive spectrometers, choppers have to be employed to modulate the incident infrared radiation.

Fourier transform infrared spectrometers are ideally suited for use with the PAS technique. No special choppers are necessary since in the FT-IR technique the interferometer acts as the modulator. Fourier transform instruments have much higher energy throughout compared to the dispersive instruments, leading to effectively larger PAS signals from the same sample when compared with dispersive instruments. The main advantage of the photoacoustic techniques lies in the fact that it enables one to obtain spectra similar to the absorbance spectra on any type of solid or semi-solid material. The samples could be solid (i.e., chunks, powders, smears, gels), crystalline, or amorphous. The PAS signal is produced only at those frequencies where the sample absorbs radiation yielding spectra with large dynamic ranges. Any radiation that is scattered, which may pose serious problems in other sampling techniques, does not contribute to the PAS technique, and this can be a major advantage in studying spectra of materials such as coals and minerals. Farrow *et al.* (1978) have used this Fourier transform PAS technique in the visible. Busse and Bullemer (1978a,b) have used the FT-IR PAS technique in the infrared. They have used a special PAS cell of a resonant type for studying the ir spectra of gases such as methanol. In Fourier transform spectroscopy, assuming a constant mirror velocity, each modulation frequency corresponds to a specific frequency in the electromagnetic spectrum. Thus, if a gas cell containing a microphone is designed to have an acoustic resonant frequency corresponding to the modulation frequency where this gas sample absorbs, a large and selective PAS signal could be recorded. Such special resonant PAS cells are known as spectrophone cells. However, these cells will not have uniform response over this entire spectral region.

For a PAS cell to be used over the entire mid-infrared region, the design criterion should be to produce a totally nonresonant cell. Rockley (1979, 1980) was the first to design a PAS cell for use with FT-IR instruments in the mid-ir range. He has examined the FT-IR PAS spectra of a wide range of materials. Rockley and Devlin (1980) have shown that the PAS technique can be very useful in studying the surface oxidation effects in coals.

Some of the practical considerations involved in the use of the PAS technique in the infrared range using Fourier transform instruments will be outlined next. Most rapid-scanning Fourier transform spectrometers that use DTGS detectors employ a mirror velocity of ~ 0.16 cm sec^{-1}. This results in modulation frequencies ranging from 0 to around 1.25 kHz over the spectral region from 0 to 4000 cm^{-1}.

Since the PAS signal is inversely proportional to the modulation frequency, the PAS signals will be lower at the high-frequency end of the mid-ir range. This is not a major problem if the spectrometer system has a high-energy throughput. Reduction of mirror velocity can improve the high-frequency sensitivity but does not alter the relative frequency response in a truly nonresonanant cell. It is also possible that when the mirror velocity is too low, corresponding to very low-modulation frequencies, the PAS cell may become very sensitive to external, low-frequency vibrations. The cells designed for use in the mid-ir range should have as little dead volume as possible.

There are two commercial PAS cells available (EG&G and Guilford Instruments) that have been used with commercial FT-IR instruments. Individual researchers have designed their own PAS cells for use with FT-IR instruments (Rockley, 1979), using resonant or nonresonant designs.

As with any other spectroscopic technique, the instrumental background has to be removed from the recorded PAS spectrum. Activated charcoal powder has been shown to be a good reference material for eliminating the instrumental background (Rockley, 1979, 1980). The activated charcoal has uniform absorbance in the mid-ir range with no sharp features discernible. When the charcoal is used as the sample in the PAS cell, the signal produced will essentially define the instrumental profile, and this spectrum can be used as the reference spectrum for eliminating the instrumental artifacts. Figure 31 shows the typical charcoal PAS spectrum from a photoacoustic detector recorded using the Digilab FTS-15 spectrometer.

As was mentioned before, the sample preparation for the photoacoustic technique is minimal; even chunks of solids can be examined. Figure 32 shows the FT-IR PAS spectrum of a piece of urethane foam recorded in 1 min. Figure 33 shows the spectrum of three different chunks of coals; Fig. 34 shows the same chunk of coal recorded after the samples had been exposed to air for a few days (Rockley and Devlin, 1980). One can clearly see the surface oxidation effects on coal from these spectra. These results also clearly indicate that the PAS technique can be very useful in surface studies—the larger the surface area, the greater will be the PAS signal. Thus, from a given sample, greater PAS signal can be produced if the sample is in the form of powder. All these spectra were corrected using reference charcoal.

Figure 35 shows a moderately high-resolution (0.5 cm^{-1}) spectrum of

Fig. 31. Reference spectrum obtained by using activated charcoal. Spectral features due to the trace amounts of the water vapor and carbon dioxide in the cell can be noted easily. The spectrum was recorded at 4 cm^{-1} resolution.

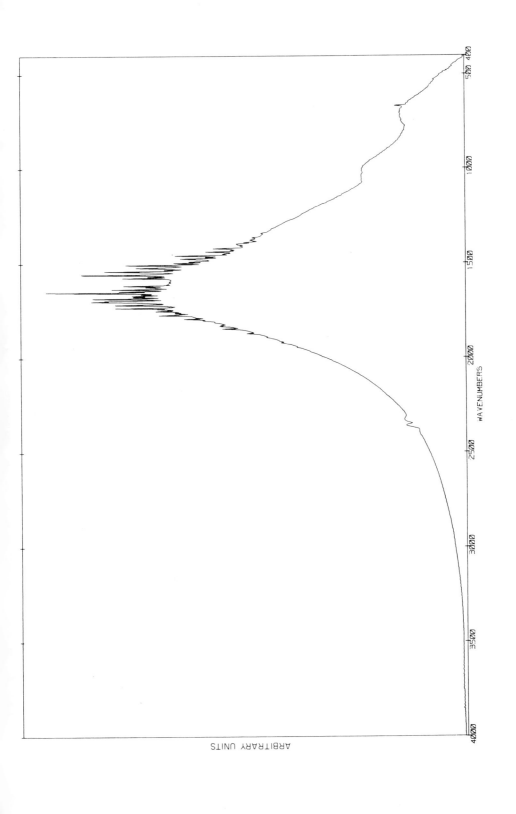

ARBITRARY UNITS

WAVENUMBERS

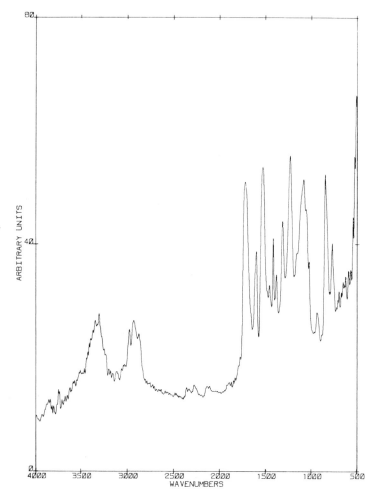

Fig. 32. Photoacoustic FT-IR spectrum of a piece of urethane foam, referenced against the charcoal. Spectrum was recorded at 8 cm^{-1} resolution with a measurement time ~1 min.

methane gas (with trace amounts of ethylene as an impurity) at atmospheric pressure. The cell contained only the gas and the spectrum is shown is that of the uncorrected single beam. Figure 36 shows an expanded plot of the C–H stretching region in the methane spectrum.

Recently, Fourier transform PAS has been extended into the biological field (Rockley et al., 1980). Spectra of hemin, hemoglobin, protopor-

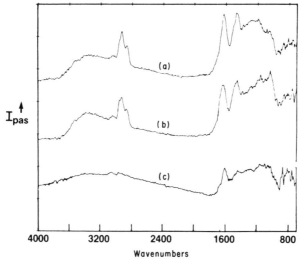

Fig. 33.   Photoacoustic FT-IR spectra of three freshly cleaved coal chunks: (a) Illinois No. 6; (b)Pittsburgh bituminous; (c) Reading anthracite (Rockley and Devlin, 1980)

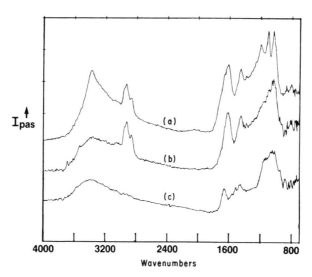

Fig. 34.   Photoacoustic FT-IR spectra of the same three chunks of coal after they have been aged for a few days. One can see the extra features due to the surface oxidation when this figure is compared with the details in Fig. 33 (Rockley and Devlin, 1980).

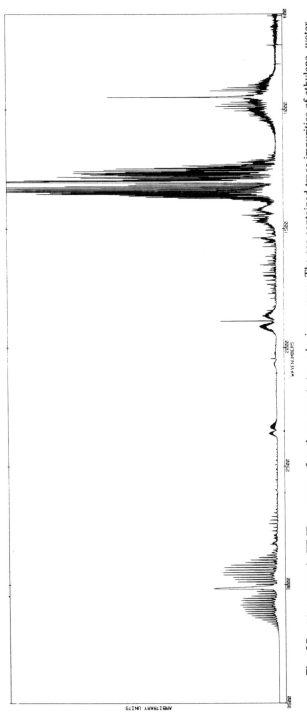

Fig. 35. Photoacoustic FT-IR spectrum of methane gas at atmospheric pressure. The gas contained trace impurities of ethylene, water vapor, and carbon dioxide. The spectrum was recorded at 0.5 cm^{-1} resolution, measurement time being 15 min. The figure shows the raw, unreferenced, single-beam spectrum.

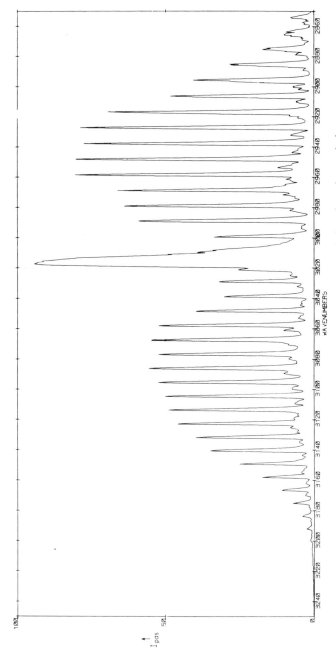

Fig. 36. The C–H stretching region of Fig. 35 scale-expanded to show the resolution.

187

phyrin IX, and horseradish peroxidose were obtained using milligram quantities of intact samples.

The preceding examples show the ease with which the infrared spectra of a variety of samples can be recorded using the photoacoustic technique. For a more expanded discussion in FT-IR PAS, see Chapter 4.

9. Gas Chromatography and Fourier Transform
   Infrared Spectroscopy

The study of gas chromatographic effluents by Fourier transform infrared spectroscopy (GC-FT-IR) is another experimental technique that is gaining acceptance rapidly. The principle of the technique is very simple. The effluent from the GC column is passed through a heated light pipe that is placed in the path of the infrared radiation from an FT-IR instrument and the infrared absorption spectra of the various effluents are recorded as they pass through the light pipe on the fly. The GC-FT-IR technique has been reviewed by Griffiths (1978) and Erickson (1979). There have been a number of papers published in the literature on the GC-FT-IR technique using packed columns (Wall and Mantz, 1977; Mattson and Julian, 1979; Krishnan *et al.*, 1979; Cottey *et al.*, 1978). Azarraga (Azarraga and Macall, 1974a,b; Azarraga, 1975), one of the pioneers in the field of GC-FT-IR, has utilized support-coated open-tubular (SCOT) glass capillary columns in the measurements. Recently Shafer *et al.* (1980) obtained the GC-FT-IR spectra using 0.5-mm glass capillary columns. Jacobs *et al.* (1980) have routinely been using wide-bore (0.75-mm internal diameter) glass capillary columns in their GC-FT-IR work. Some of the considerations involved in the design of special light pipes for different columns and other parameters involved in the design of the GC-FT-IR systems have been discussed by Erickson (1979).

The GC-FT-IR light pipe is, typically, a 1- or 2-mm internal diameter, 40- or 60-cm-long, gold-coated glass capillary tubing. The transmission of the light pipe is typically of the order of 5 to 10%, necessitating the use of liquid-nitrogen-cooled MCT detectors. The light pipe is connected to the GC column by means of suitable heated transfer lines. Some light pipes (such as the one offered by Digilab) contain air-actuated pneumatic valves that can be used to trap specific GC fractions in the light pipe for improving the sensitivity of the measurements by collecting data for longer periods. If the GC contains a nondestructive detector such as a thermal conductivity detector, the GC effluents coming out of the GC detector can be directed into the FT-IR light pipe. However, if the GC uses a destructive detector such as a flame ionization detector, then an effluent splitter has to be used. This splitter is placed between the GC column and the GC detector and is configured to direct most of the effluents from the GC onto the light pipe.

The GC-FT-IR technology is fairly well established by now, with FT-IR spectra of GC fractions corresponding to as little as 50 ng being recorded routinely. Figure 37 shows the GC-FT-IR spectrum of 60 ng of isobutylmethacrylate recorded using Digilab FT-IR instrumentation in conjunction with a packed GC column. In performing the GC-FT-IR experiments, it is good to bear in mind that the best spectrum for a given GC fraction can be obtained when all of the fraction can be directed through the light pipe at one time. This means that the conditions for the GC separation must be adjusted such that each GC peak (as seen by the GC detector) is as sharp as possible.

The GC-FT-IR technique is capable of enhancing the resolution of the GC column. If there is a gas chromatographic peak that contains, for instance, two unresolved components, GC-FT-IR data can be collected as different parts of the GC fraction (namely the leading edge, the trailing edge, and so on) are passing through the light pipe. This procedure would result in a number of GC-FT-IR spectra, and the use of the absorbance-subtraction technique will allow one to obtain the spectra of the pure com-

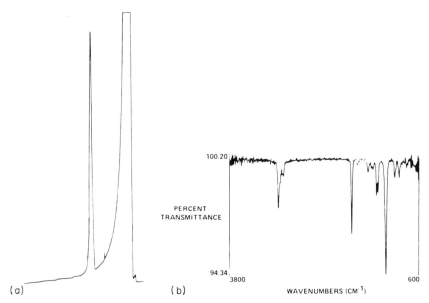

Fig. 37.   (a) Chromatogram and (b) unsmoothed spectrum showing GC-FT-IR sensitivity of 60 ng of isobutylmethacrylate on the fly (Wall and Mantz, 1977).

ponents. Alternately, GC-FT-IR data can be collected over the GC fraction of interest with a fixed time resolution. This will result in a number of
spectra profiling the passage of the GC fraction through the light pipe.
Then, functional group chromatograms (called *infragrams,* or *chemi-
grams*) can be obtained by monitoring the infrared absorbances over specific frequency windows or frequencies in these spectra and thus enhnace
the GC resolution. An example of such a procedure for enhancing the GC
resolution obtained for a mixture of chlorotoluene isomers is shown in
Fig. 38. Figure 39 shows the GC-FT-IR spectra from this mixture. Note
that a packed column was used in these experiments (Krishnan *et al.,*
1979).

Most of the GC-FT-IR light pipes that are commonly used may contain
some metal parts. Such light pipes cannot be used with materials such as
natural products and pesticides, which may decompose on contact with
metals. In such cases, all-glass GC-FT-IR accessories need to be used in
conjunction with glass-lined transfer lines. Figure 40 shows the GC-FT-IR
spectrum of the pesticide parathion. Figure 41 shows the GC-FT-IR spectrum of the drug sec-amylbarbitol recorded using a Digilab all-glass GC-
FT-IR system. In the latter case, it was found that even the thermal conductivity detector in the GC was sufficient to decompose the material, and

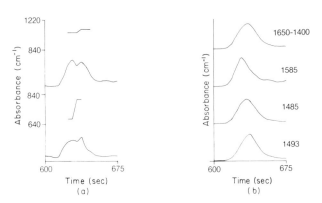

Fig. 38. Reconstructed chromatograms from a mixture of chlorotoluene isomers. (a)
Functional group chromatograms monitoring the absorbance between 840 and 1220 cm^{-1}
and between 640 and 840 cm^{-1}. The chromatogram was sampled by the FT-IR in successive
1.5-sec time resolution slices. The figure shows the absorbance as a function of time in the
window regions, as well as the frequencies of the major bands causing the absorption. (b)
Reconstructed chromatograms over the window 1650–1400 cm^{-1} and at fixed frequencies
indicated. In the present example the functional group chromatogram is not able to fully
resolve the three chlorotoluene isomers; the specific frequency chromatograms are able to
sense the onset of each of the isomers. Reproduced from Krishnan *et al.* (1979), by permission of Preston Publications, Inc.

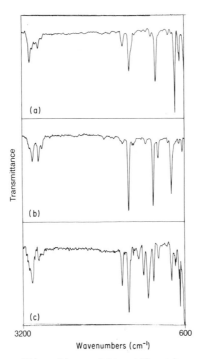

Fig. 39. The spectra of (a) *o*-, (b) *p*-, and (c) *m*- chlorotoluenes (in the order of elution) obtained from the data shown in Fig. 38. Reproduced from Krishnan *et al.* (1979), by permission of Preston Publications, Inc.

had to be disconnected before the spectrum shown in Fig. 41 could be recorded.

The current trends in GC-FT-IR technology are toward improved capillary GC-FT-IR performance and improved software. The use of high-speed array processors in the FT-IR data systems allows real-time GC-FT-IR spectroscopy to be possible. (Interferograms can be computed as fast as they are collected so that at the end of a GC run, one has the spectra all computed.) Software techniques, such as the Gram–Schmidt orthogonalization procedure developed by Isenhour and co-workers (Hanna *et al.*, 1979), can allow the GC-FT-IR accessory to be used as an infrared detector for the GC, indicating all the fractions in a given GC mixture that are likely to yield infrared spectra.

In fact, the collection of data over the GC fractions, the Fourier transform of and the creation of absorbance spectra from the collected interferograms, the creation of either the functional group or Gram–Schmidt reconstructed chromatograms along with the real-time display of the col-

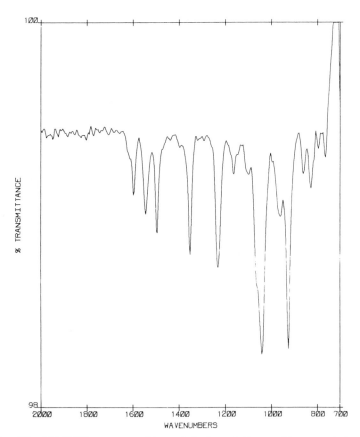

Fig. 40.   GC-FT-IR spectrum of 1 μg of the insecticide parathion recorded using an all-glass system.

lected full absorption spectra can all be performed at one time by the cur-rent-generation GC-FT-IR software. Vapor-phase spectral libraries are being developed for use with search routines for the identification of un-known GC-FT-IR spectra. Some of these recent developments are illus-trated in Figs. 42 and 43. Figure 42 shows the chromatograms and differ-ent reconstructed chromatograms from a multicomponent mixture, recorded using a wide-bore (0.75-mm) glass capillary column. Figure 43 shows some selected GC-FT-IR spectra from this mixture and the output of the corresponding spectral search results.

B.   Computer Techniques

1.   Fringe Removal

Thin plane parallel samples in spectroscopy often show channel spectra due to multiple reflections within the sample. The channel spectrum manifests itself as interference fringes throughout the spectrum. These fringes can often obscure the weaker features in the spectrum. The problem can be overcome to a limited extent by preparing samples with matted surfaces by crumpling the film or by putting a hot iron over the sample when it is placed over a rough surface.

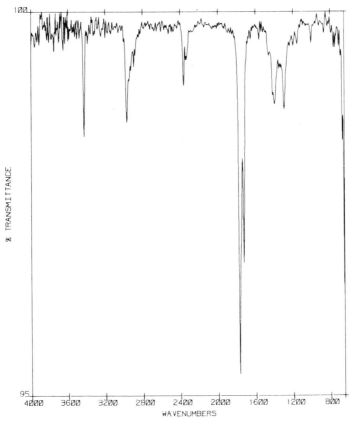

Fig. 41.   The GC-FT-IR spectrum of the drug sec-amylbarbitol recorded using an all-glass system.

The interference fringes produced from the plane parallel sample will have their fringe maxima located at wavenumbers

$$\nu_m = m/2nd$$

where m = 0, 1, 2, 3, . . . is the order of interference; n the sample refractive index; and d its thickness. The intensity at these fringe maxima will be given by

$$\triangle T = 4(n - 1)^2/(n + 1)^2$$

For samples that are highly transparent, it can be reasonably assumed that the refractive index is constant throughout a large spectral region, such as the mid-infrared region. These fringes will then be a constant frequency sinusoid in the wavenumber space. In the Fourier space, the fringes will then manifest themselves as a secondary maximum or side-

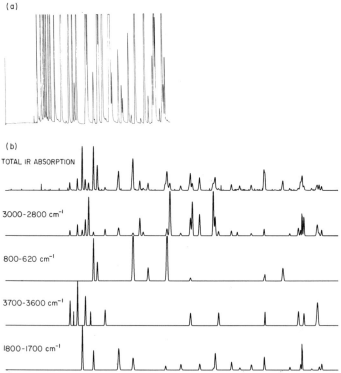

Fig. 42. (a) The chromatogram and (b) various infrared reconstructed chromatograms from a very complex GC mixture. The total ir absorption reconstructed chromatogram refers to the Gram–Schmidt reconstruction from the interferogram space. Reprinted from *Am. Lab.* **13**(3), 122 (1981). Copyright 1981 by International Scientific Communications, Inc.

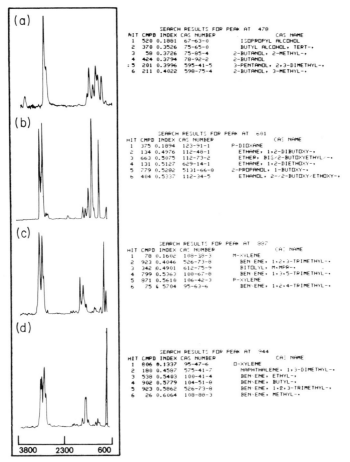

Fig. 43. GC-FT-IR spectra and the results of a spectral library search on some of the GC fractions from Fig. 42: (a) isopropyl alcohol, 1050 V; (b) p-dioxane, 506 V; (c) m-xylene, 139 V; (d) o-xylene, 1579 V. The spectral search results are presented with a quality index that represents the difference between the spectrum being searched and the library spectrum. Reprinted from *Am. Lab.* 13(3), 122 (1981). Copyright 1981 by International Scientific Communications, Inc.

burst. Hirschfeld and Mantz (1976) have described a software technique for removing this sideburst from the interferogram. If the sideburst is well removed from the main peak or the centerburst of the interferogram and the sideburst is very narrow, then, by suitable software routines, the sideburst can be replaced by zeros. However, by this procedure, some real spectral information contained in the region of the sideburst would also have been removed and this could lead to a very slight distortion of the subsequent transformed spectrum. This distortion is due mainly to the phase errors introduced by the procedure. A better method for removing the sideburst, described by Hirschfeld and Mantz (1976), involves the use of two sample interferograms. These are recorded for two orientations of the samples such that the optical thicknesses of the sample are different in the two cases. The sideburst in the two interferograms will then occur at different locations. A FORTRAN program can then be used to replace the sideburst in one of the interferograms with the information from the same locations in the second interferogram.

These procedures would be very useful for eliminating the fringes from the spectra of moderately thick samples where the sideburst could easily be located in the interferograms of the samples. However, when the samples are very thin, the sideburst may occur very close to the centerburst, making them not very easily identifiable. In the following sections a variation of these procedure for eliminating the fringes will be described.

Figure 44 shows two interferograms of a very thin polystyrene film. The two interferograms have been recorded for two different orientations of the sample. However, the sideburst could not be easily identified in

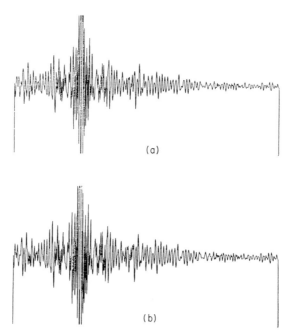

Fig. 44. Interferograms of a polystyrene film recorded for two sample orientations relative to the ir beam: (a) position 1; (b) position 2.

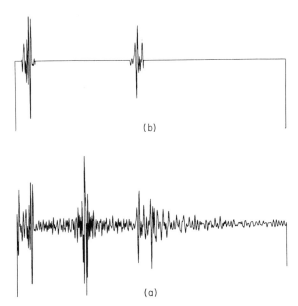

Fig. 45.   (a) Difference between the two interferograms shown in Fig. 43, showing two sidebursts; (b) after zeroing all feature except sidebursts corresponding to one thickness.

these interferograms. A difference interferogram between the two could be generated and is shown in the Fig. 45a. The scale in this figure has been expanded by a factor of ten in comparison to Fig. 44. In the difference interferogram one can see a trace of the uncanceled centerburst and two sidebursts to the right of it. The two sidebursts obviously correspond to the two sample thicknesses. A software routine can be used to replace the data in the interferogram with zeros, except in two specified window regions. We could then use this program to replace all the features in the difference interferogram except for the sideburst arising from one sample thickness. The result of this operation is shown in Fig. 45b. This zeroed interferogram therefore contains essentially only the information about the fringes due, in the present example, from the smaller of the two sample thicknesses. The zeroed interferogram could then be subtracted from the original sample interferogram, corresponding the lower thicknesses, and the resulting transformed spectrum should be free, to a large extent, of the fringes. This can be seen in Fig. 46, where trace (a) is the actual spectrum of the sample and trace (b) is the spectrum corrected using the procedure just outlined. One can see from the figure that the fringes have completely been eliminated.

Another method of fringe removal, by subtracting the electronically

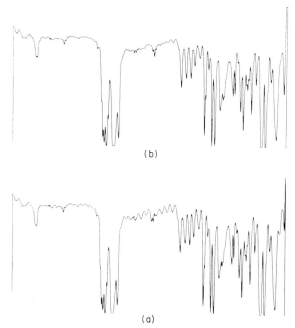

Fig. 46.   (a) Original spectrum; (b) fringe-removed spectrum.

generated sine wave of the appropriate frequency from the spectrum containing fringes, has also been described recently (Clark and Moffat, 1978).

2.   Absorbance-Subtraction Technique

The difference, or absorbance-subtraction, spectroscopy provides a very sensitive method for detecting small differences between samples. These differences could be differences due to the physical state of the sample, for example, crystalline and amorphous polymer; or the difference could be a small amount of the solute dissolved in a solvent. Common features in the spectra being compared cancel out and the remaining bands can be interpreted in terms of the differences between the samples. Prior to the advent of the FT-IR instruments and computerized data handling, only the dual-beam ratio technique could be used to obtain difference spectra. If, for instance, the spectrum of a solute in a solution is required, matched pathlength cells—one containing the solution and the other containing the pure solvent—should be placed in the two beams of a dual-beam dispersive instrument. Assuming that the solute concentration is low enough, the resulting spectrum would only be that of the solute. This technique is severely limited in use since only two samples at a time can be compared, and low concentrations and matched pathlengths are needed.

With the advent of the FT-IR instruments and computerized data-handling capabilities, the absorbance-subtraction techniques has become very popular. If two samples are to be compared, their absorbance spectra are created and stored on the data system. Difference spectra can then be generated by canceling out common bands in the two spectra. Since, according to the Beer–Lambert law, the absorbances are linearly proportional to pathlength or concentration, any differences between the two spectra being compared, in pathlength or concentration, can be taken into account by using suitable scaling factors. Thus the use of exactly matched pathlength cells is no longer necessary. Any number of spectra, taken in pairs, can be compared, thus making the analysis of multicomponent mixtures possible. The principles involved and some of the applications of the absorbance-subtraction technique have been discussed by Koenig (1975).

It has been observed experimentally that the absorbance-subtraction technique works very well only when the absorbances are less than 0.6–0.8 absorbance units. That is, if one is looking for a solute band buried under a solvent band, the solvent band could be cleanly subtracted away to clearly show the presence of the solute band, only if the peak absorbance of the solvent band is less than 0.6–0.8 absorbance units. If the absorbances are stronger, then the bands do not cancel out exactly, leading to the presence of artifacts in the difference spectrum. Thus, if a meaningful difference spectrum over the entire spectral range is needed, the sample preparation should be such that the maximum peak absorbances in the spectra are in the range just specified.

The absorbance-subtraction technique can also provide useful information about chemical or physical changes in the samples. If a solution and the pure solvent spectra are being compared, the difference spectrum can yield information about any solvent–solute interactions. Such interactions will usually result in band-broadening or frequency shifts between the solution and the pure solvent spectra. If a particular spectral band has undergone broadening as a result of solvent–solute interactions, the difference spectrum will have this shape of a double derivative; if there has been a frequency shift, the difference spectrum will appear as a first derivative. These effects are illustrated in Fig. 47 on a synthetic Lorentzian peak. Improper use of the subtraction technique, when such solvent–solute interactions exist, will result in artificial features being present in the difference spectrum.

The absorbance-difference technique can also be used for qualitative and quantitative analysis of multi-component mixtures. As was pre-

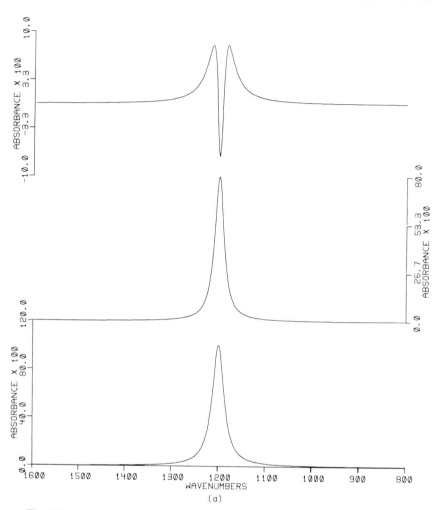

Fig. 47. Subtraction between synthesized Lorentzian bands showing the effects of (a) band broadening and (b) frequency shift leading to artifacts in the difference spectrum.

viously discussed, too strong absorbances and solvent–solute interactions can affect the qualitative features in the difference spectrum. There are a number of other factors that might affect the quantitative accuracy in the measurements. Some of these have been discussed in detail by Hirschfeld (1979). Anderson and Griffiths (1975) have shown that substantial deviations from Beer's law occur when the spectral resolution employed in the measurements is comparable to the bandwidths of the spectral lines being measured. Ramsey's criterion states that for the

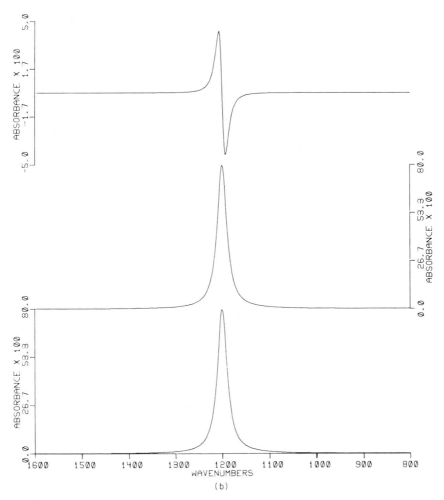

Fig. 47 (*Continued*)

measurement of absorbances, the resolution employed should be at least one-fifth that of the bandwidths. In FT-IR measurements, a somewhat more stringent form of the Rayleigh criterion may be needed. Griffiths has also indicated that the use of the boxcar as opposed to the triangular apodization in the Fourier transform calculations will improve quantitative accuracy. This effect, however, may not be very substantial. In general, since most liquids and solids have spectral bandwidths in the range of 10 to 30 cm^{-1} at room temperature, the use of 2 or 1 cm^{-1} spectral resolution will easily overcome any resolution errors just discussed.

As had been discussed in a separate section, the presence of interference fringes arising from plane parallel samples can affect the difference spectrum. Strongly wedged samples (Hirschfeld, 1979) can also affect the difference spectrum. This would be the case in, say, the spectra of viscous samples run as smears on salt plates. If such samples are of nonuniform thickness across the infrared beam, different parts of the beam would go through different thicknesses of the sample. This would result in distortions of the band shapes, and the resulting difference spectrum would contain artificial features. For such samples, the use of films cast from some solution rather than smears on salt plates would go a long way

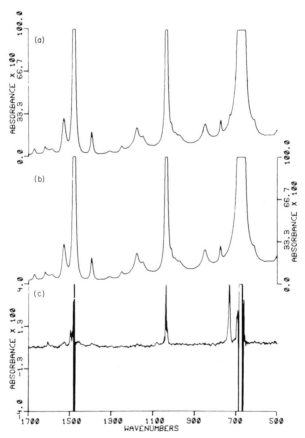

Fig. 48. Absorbance-subtraction brings out the presence of the 735 cm^{-1} band of toluene from the wing of the strong benzene band: (a) spectrum of 0.1% toluene in benzene; (b) spectrum of pure benzene; and (c) difference spectrum showing the toluene bands.

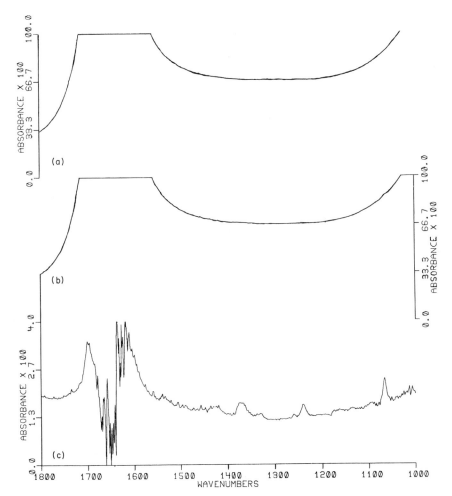

Fig. 49. Subtraction process showing: (a) the spectrum of 0.2% acetone in water; (b) the spectrum of pure water; and (c) the difference spectrum. A spectral resolution of 2 cm⁻¹ and triangular apodization were employed in the measurements.

toward overcoming the wedging problem. Another way of overcoming the wedging error would be to use apertures in front of the sample.

The absorbance-subtraction technique has been used in quality control, in compound identification, and for the study of surface effects (Manocha and Montgomery, 1978; Lowry and Banzer, 1978). Some of the spectral artifacts in the difference spectra and the conditions that might cause these effects have also been studied (Strassburger and Smith, 1979).

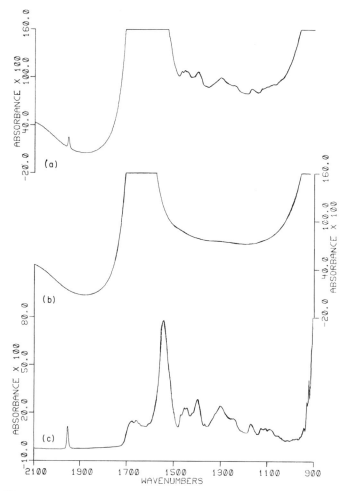

Fig. 50. Subtraction process showing the spectrum of human blood exposed to carbon monoxide. A resolution of 2 cm⁻¹ and triangular apodization were used. (a) Spectrum of the blood; (b) reference water; and (c) difference sectrum.

The absorbance-subtraction technique has also been applied to the study of geometrical (Gendreau *et al.*, 1976) and stereoisomers (Gendreau and Griffiths, 1976). Koenig and his co-workers have made extensive use of the absorbance-subtraction technique. They have used the technique for studying the irradiation damage in polyethylene (Tabb *et al.*, 1975) and for obtaining the spectra of crystalline and amorphous polymers (Coleman *et al.*, 1974; Tabb and Koenig, 1975). Lin *et al.* (1979) have used the tech-

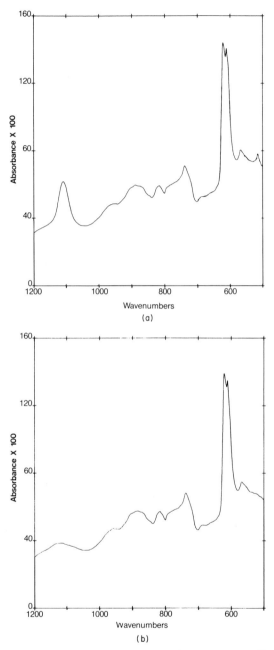

Fig. 51. Application of the absorbance-subtraction process to silicon. Spectra, recorded at 1 cm^{-1} resolution and triangular apodization, of: (a) a 2-mm-thick Czochlorski-grown silicon wafer at 300°K; (b) a 2-mm-thick float-zone-grown silicon reference wafer free of oxygen and carbon at 300°K; and (c) the difference spectrum showing the oxygen (1106 and 520 cm^{-1}) and the carbon (607 cm^{-1}) bands at 300°K.

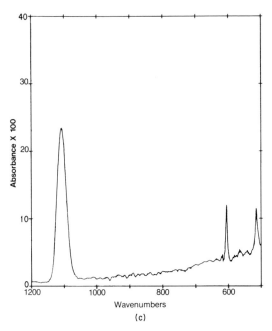

Fig. 51   (*Continued*)

nique for studying the thermal and photochemical changes in polymers. The technique has been used in conjunction with polarization measurements to enhance the resolution in the infrared spectrum (Krishnan, 1978).

The absorbance-subtraction technique is capable of resolving even very weak features due to, say, a solute buried under strong absorption bands due to the solvent. This is illustrated in Fig. 48, which shows the subtraction process clearly bringing out the 735 cm^{-1} toluene band buried in the shoulder of a much stronger absorbance due to benzene, in a solution of 0.1% toluene in benzene. Figure 49 shows a similar result from a solution of 0.2% acetone in water. In spite of the fact that the water band at ~ 1650 cm^{-1} is very strong, the subtraction clearly brings out the carbonyl band from acetone. Figure 50 shows the absorbance-subtraction process applied to human blood. The technique has shown its power and usefulness particularly in aqueous systems, so the study of the infrared spectra of aqueous systems is not as difficult as it used to be with dispersive instruments. Another application of the absorbance-subtraction technique is in the semiconductor industry where it is used for monitoring the oxygen and carbon contents of silicon and germanium wafers. Figure 51 shows the results of oxygen and carbon measurements on a 2-mm-thick Czochlorski-grown silicon wafer, using a float-zone-refined reference wafer. The absorbance-subtraction technique is used routinely nowadays in conjunction with every spectroscopic technique, such as ATR, diffuse reflectance, GC-FT-IR, and LC-FT-IR.

REFERENCES

Adams, D. M., and Payne, S. J. (1972). *In* "Annual Reports A," pp. 3–17. The Chemical Society, London.

Adams, D. M., and Sharma, S. K. (1977). *J. Phys. E* **10**, 838.

Anderson, D. H., Wilson, T. E. (1975). *Anal. Chem.* **47**, 2482.

Anderson, R. J., and Griffiths, P. R. (1975). *Anal. Chem.* **47**, 2339.

Azarraga, L. V. (1975). Presented at the *Ann. Symp. Recent Adv. Anal. Chem. Pollutants, 5th, Jekyll Island, Georgia, May*.

Azarraga, L. V., and McCall, A. C. (1974a). *In* "Infrared Fourier Transform Spectrometry of Gas Chromatographic Effluents," EPA-660/2-73-034, p. 61.

Azarraga, L. V., and McCall, A. C. (1974b). Presented at *Pittsburg Conf. Anal. Chem. Appl. Spectrosc., 25th, Cleveland, Ohio, March*.

Barr, J. K. (1969). *Infrared Phys.* **9**, 97.

Bates, J. B. (1978). *In* "Fourier Transform Infrared Spectroscopy" (J. R. Ferraro and L. J. Basile, eds.), Vol. 1, p. 99. Academic Press, New York.

Bates, J. B., and Boyd, G. E. (1973). *Appl. Spectrosc.* **27**, 204.

Bell, A. G. (1881). *Phil. Mag.* **11**, 510.

Billmeyer, F. W. Jr., and Saltzman, M. (1966). "Principles of Color Technology." Wiley, New York.

Blevin, W. R., and Brown, W. J. (1965). *J. Sci. Instrum.* **42**, 385.

Brasch, J. W. (1965). *J. Chem. Phys.* **43**, 3473.

Busse, G., and Bullemer, B. (1978a). *Infrared Phys.* **18**, 255.

Busse, G., and Bullemer, B. (1978b). *Infrared Phys.* **18**, 631.

Clark, F. R. S., and Moffat, D. J. (1978). *Appl. Spectrosc.* **32**, 547.

Coblentz, W. W. (1913). *Nat. Bur. Std. Bull.* **9**, 283.

Coleman, M. M., Painter, P. C., Tabb, D. L., and Koenig, J. L. (1974). *Polym. Lett.* **12**, 577.

Cottey, P., Mattson, D. R., and Wright, J. C. (1978). *Am. Lab.* **10**, 126.

Cournoyer, R., Sheayer, J. C., and Anderson, D. H. (1977). *Anal. Chem.* **49**.

Dunn, S. T. (1965). Flux Averaging Devices in the Infrared. Nat. Bur. Std. Tech. Note No. 279.

Dunn, S. T., Richmond, J. C., and Wiebelt, J. A. (1966). *J. Res. Nat. Bur. Std. Sect C* **70**, 75.

Durig, J. R., and Cox, A. W. Jr. (1978). *In* "Fourier Transform Infrared Spectroscopy." (J. R. Ferraro and L. J. Basile, eds.), Vol. 1, p. 215. Academic Press, New York.

Erickson, M. D. (1979). *Appl. Spectrosc. Rev.* **15**, 261.

Farrow, M. M., Burnham, R. K., and Eyring, E. M. (1978). *Appl. Phys. Lett.* **33**, 735.

Ferraro, J. R., Mitra, S. S., and Postmus, C. (1966). *Inorg. Nucl. Chem. Lett.* **2**, 269.

Ferraro, J. R. (1971). *In* "Spectroscopy in Inorganic Chemistry" (C. N. R. Rao and J. R. Ferraro, eds.), pp. 55–77. Academic Press, New York.

Ferraro, J. R., and Basile, L. J. (1974). *Appl. Spectrosc.* **28**, 505.

Ferraro, J. R., and Basile, L. J. (eds.) (1978). "Fourier Transform Infrared Spectroscopy," Vol. 1. Academic Press, New York.

Ferraro, J. R., and Basile, L. J. (eds) (1979a). "Fourier Transform Infrared Spectroscopy," Vol. 2. Academic Press, New York.

Ferraro, J. R., and Basile, L. J. (1979b). *Am. Lab.* **11**, 31.

Ferraro, J. R., and Basile, L. J. (1980). *Appl. Spectrosc.* **34**, 217.

Ferraro, J. R., Walling, P. L., and Sherron, A. T. (1980). *Appl. Spectrosc.* **34**, 570.

Francis, S. A., and Ellison, A. H. (1959). *J. Opt. Soc. Am.* **49**, 131.

Frei, R. W. and McNeil, J.D. (1973). "Diffuse Reflectance Spectroscopy in Environmental Problem Solving." CRC Press, Cleveland, Ohio.

Fringeli, U. P. (1977). *Z. Naturforsch.* **32c**, 20.

Fuller, M. P., and Griffiths, P. R. (1978a). *Anal. Chem.* **50**, 1906.

Fuller, M. P., and Griffiths, P. R. (1978b). *Am. Lab.* **10**, 69.

Gendreau, R. M., and Griffiths, P. R. (1976). *Anal. Chem.* **48**, 1910.

Gendreau, R. M., Griffiths, P. R., Ellis, L. E., and Anfinsen, J. R. (1976). *Anal. Chem.* **48**, 1907.

Gier, J. T., Dunkle, R. V., and Bevans, J. T. (1954). *J. Opt. Soc. Am.* **44**, 558.

Greenler, R. G. (1966). *J. Chem. Phys.* **44**, 310.

Griffiths, P. R. (1975). *Am. Lab.* **7**, 37.

Griffiths, P. R. (1978). *In* "Fourier Transform Infrared Spectroscopy" (J. R. Ferraro and L. J. Basile, eds.), Vol. 1, pp. 143–168. Academic Press, New York.

Hadni, A. (1967). "Essentials of Modern Physics Applied to the Study of the Infrared," pp. 134–246. Pergamon, Oxford.

Handke, M., Stoch, A., Lorenzelli, V., and Bonora, P. L. (1980). *J. Mater. Sci.* **15**, 1317.

Hanna, D. A., Hangac, G., Hohne, B. A., Small, G. W., Wiebolt, R. C., and Isenhour, T. L. (1979). *J. Chromat. Sci.* **17**, 423.

Hardy, A. C. (1936). "Handbook of Calorimetry." MIT Press, Cambridge, Massachusetts.

Harrick, N. J. (1967). "Internal Reflection Spectroscopy." Wiley (Interscience), New York.

Hirschfeld, T. (1979). "Fourier Transform Infrared Spectroscopy" (J. R. Ferraro and L. J. Basile, eds.), Vol. 2, p. 193. Academic Press, New York.

Hirschfeld, T., and Mantz, A. W. (1976). *Appl. Spectrosc.* **30**, 552.

Ishitani, A., Ishida, H., Soeda, F., and Nagasawa, Y. (1980). *Pittsburg Conf. Anal. Chem. Appl. Spectrosc.*

Jacobs, M., Ledig, W., and Waldrat, J. (1980). Presented at the *Digilab User's Conf., Cambridge, Massachusetts, June.*

Jacobsen, R. J., Mikawa, Y., and Braszh, J. W. (1970). *Appl. Spectrosc.* **24**, 33.

Jacobsen, R. J. (1979). *In* "Fourier Transform Infrared Spectroscopy" (J. R. Ferraro and L. J. Basile, eds.), Vol. 2, p. 165. Academic Press, New York.

Judd, D. B. and Wyszecki (1963). "Color in Business, Science and Industry," 2nd ed. Wiley, New York.

Kanstad, S. O., and Nordal, P.-E. (1977). *Int. J. Quant. Chem. Suppl. 2* **12**, 123.

Kanstad, S. O., and Nordal, P.-E. (1979). *Infrared Phys.* **19**, 413.

Kember, D., and Sheppard, N. (1975). *Appl. Spectrosc.* **29**, 496.

Kember, D., Chenery, D. H., Sheppard, N., and Fell, J. (1979). *Spectrochim. Acta* **35A**, 455.

Klaeboe, P., and Woldback, T. (1978). *Appl. Spectrosc.* **32**, 588.

Koenig, J. L. (1975). *Appl. Spectrosc.* **29**, 293.

Kortüm, G. (1969). "Reflectance Spectroscopy." Springer-Verlag, Berlin and New York.

Kortüm, G., and Delfs, H. (1964). *Spectrochim. Acta* **20**, 405.

Kortum, P., Braun, W., and Herzog, G. (1963). *Z. Angew. Chem. Int. Ed. Engl.* **2**, 333.

Krishnan, K. (1978). *Appl. Spectrosc.* **32**, 549.

Krishnan, K., Curbelo, R., Chiha, P., and Noonan, R. C. (1979). *J. Chromat. Sci.* **17**, 413.

Krishnan, K., Hill, S. L., and Brown, R. H. (1980). *Am. Lab.* **12**, 104.
Kubelka, P. (1948). *J. Opt. Soc. Am.* **38**, 448.
Kubelka, P., and Munk, F. (1931). *Z. Tech. Phys.* **12**, 593.
Lacy, M. E. (1979). *Contaminat. Control Seminar, Anaheim, California, October 16–19.*
Lauer, J. L. (1978). *In* "Fourier Transform Infrared Spectroscopy" (J. R. Ferraro and L. J. Basile, eds.), Vol. 1, pp. 169–213. Academic Press, New York.
Lin, S., Bulkin, B. J., and Pierce, E. (1979). *J. Polym. Sci.* **17**, 3121.
Lippincott, E. R., Weir, C. E., Van Valkenburg, H., and Bunting, E. N. (1960). *J. Res. Nat. Bur. Std. U.S. Sect. A Spectrochem. Acta* **16**, 58.
Lippincott, E. R., Welsh, F. E., and Weir, C. E. (1961). *Anal. Chem.* **33**, 137.
Low, M. J. D., and Clancy, F. K. (1967). *Environm. Sci. Technol.* **1**, 73.
Low, M. J. D., and Parodi, G. A. (1980a). *Appl. Spectrosc.* **34**, 76.
Low, M. J. D., and Parodi, G. A. (1980b). *Infrared Phys.* **20**, 333.
Lowry, S. R., and Banzer, J. D. (1978). *Anal. Chem.* **50**, 1187.
Manocha, A. S., and Montgomery, R. M. (1978). *Appl. Spectros.* **32**, 344.
Mattson, D. R., and Julian, R. J. (1979). *J. Chromat. Sci.* **17**, 416.
McDevitt, N. T., Wilkowski, R. E., and Fateley, W. C. (1967). Abstract, *Colloq. Spectrosc. Int. 13th, Ottawa, Canada, June 8–24.*
Nordal, P.-E., and Kanstad, S. O. (1977). *Int. J. Quant. Chem. Suppl. 2,* **12**, 115.
Postmus, C., Ferraro, J. R., and Mitra, S. S. (1968). *Inorg. Nucl. Chem.* **4**, 155.
Postmus, C., Nakamoto, K., and Ferraro, J., (1967). *Inorgan. Chem.* **6**, 194.
Rockley, M. G. (1979). *Chem. Phys. Lett.* **68**, 455.
Rockley, M. G. (1980). *Appl. Spectrosc.* **34**, 405.
Rockley, M. G., and Devlin, J. P. (1980). *Appl. Spectrosc.* **34**, 407.
Rockley, M. G., Davis, D. M., and Richardson, H. H. (1980). *Science* **210**, 918.
Rodgers, J. L. (1972). *High Temp-High Pressure* **4**, 271.
Röntgen, W. C. (1881). *Phil. Mag.* **11**, 308.
Rosencwaig, A. (1975). *Anal. Chem.* **47**, 592A.
Rosencwaig, A., and Gersho, A. (1976). *J. Appl. Phys.* **47**, 64.
Shafer, K. H., Bjorseth, A., Tabor, J., and Jakobsen, R. J. (1980). *J. High Res. Chrom. Commun.* **3**, 87.
Smith, F. (1980). *Digilab User's Conf.*
Strassburger, J., and Smith, I. T. (1979). *Appl. Spectrosc.* **33**, 283.
Tabb, D. L., and Koenig, J. L. (1975). *Macromolecules* **8**, 829.
Tabb, D. L., and Sevick, J. J., and Koenig, J. L. (1975). *J. Polym. Sci.* **13**, 815.
Tyndall, J. (1881). *Proc. R. Soc. London Ser. A.* **31**, 307.
Wall, D. L., and Mantz, A. W. (1977). *Appl. Spectrosc.* **31**, 525.
Weir, C. E., Lippincott, E. R., Van Valkenburg, A., and Bunting, E. N. (1959). *J. Res. Nat. Bur. Std. U.S. Sect. A* **63**, 55.
White, J. V. (1964). *J. Opt. Soc. Am.* **54**, 1332.
Willey, R. R. (1976). *Appl. Spectrosc.* **30**, 593.
Wood, B. E., Pipes, J. G., Smith, A. M., and Roux, J. A. (1976). *Appl. Opt.* **15**, 940.

INDEX